LIFE-COST APPROACH TO BUILDING EVALUATION

LIFE-COST APPROACH TO BUILDING EVALUATION

Craig A.Langston

ELSEVIER
BUTTERWORTH
HEINEMANN

AMSTERDAM BOSTON HEIDELBERG LONDON NEW YORK OXFORD
PARIS SAN DIEGO SAN FRANCISCO SINGAPORE SYDNEY TOKYO

Published throughout the world excluding Australia and New Zealand by
Butterworth Heinemann
An imprint of Elsevier
Linacre House, Jordan Hill, Oxford OX2 8DP
30 Corporate Drive, Burlington, MA 01803
www.elsevier.com

Published in Australia and New Zealand by
University of New South Wales Press Ltd
University of New South Wales
Sydney NSW 2041
AUSTRALIA

First published 2005

British Library Cataloguing in Publication Data
A catalogue record for this book is available from the British Library

Library of Congress Cataloguing in Publication Data
A catalogue record for this book is available from the Library of Congress

ISBN 0 75066 630 7

CONTENTS

PREFACE

This book is primarily designed for students who are undertaking courses in construction, property, engineering and architecture disciplines to help them understand the principles and applications of life-cost studies. It can also be used as a self-learning tool by practitioners.

It is written in a form that resembles an academic semester, with weekly content (chapters), review questions, worked examples and key points that constitute a useful teaching resource. It can be applied to face-to-face educational contexts, or as a student guide in distance learning or online delivery. A complete set of presentation slides for each chapter provides a coherent lecture series, and can be downloaded from the publisher's website (www.elsevier.com).

The material is prepared for an international audience and includes recommended reading applicable to all countries and cultures. The focus is the built environment. Concepts are deliberately presented in an understandable manner using plain English supplemented with frequent illustrations.

The style adopted is derived from many years of teaching experience to domestic and international students, regular feedback, refereed publications and research. I hope, in whatever way this book may be employed, that you find it both interesting and informative.

Professor Craig Langston
Deakin University, Australia

CHAPTER 1

A TOTAL COST APPROACH

LEARNING OBJECTIVES
During this chapter you will learn about whole-of-life considerations that underpin life-cost studies, and the revival of the technique due to sustainable development objectives. By the end of this chapter you should be able to:
- relate the importance of life-cost studies to the sustainability movement,
- understand and explain terms like life-cost, life cycle and terotechnology,
- appreciate the importance, and difficulty, of forecasting future events, and
- explain the difference between the concepts of cost and value.

CONTEXT

The environmental crisis is of increasing global importance. It is essential to recognise that the environment is affected by human economic activity and yet human economic activity in turn relies totally on the environment. There is of necessity a vital partnership between economics and nature; a balance of immense significance to the long-term prosperity of mankind.

Over the past decade there has been worldwide consensus on the need for sustainable development. Alarming realisations about the rate of depletion of our natural environment caused by the stress of economic growth have resulted in a larger proportion of the world's population becoming environmentally conscious. Attitudes towards conservation and pollution are gradually changing, but as the twenty-first century unfolds this change will, of necessity, need to occur at a faster rate.

Development also implies change, and should by definition lead to an improvement in the quality of life of individuals. Development encompasses not only growth but general utility and well-being, and involves the transformation of natural resources into productive output. Therefore the environment and the economy necessarily interact. Sustainable development is the balance between economic progress and environmental conservation, recognising that both are imperative to our future survival?

Sustainable development implies using renewable natural resources in a way that does not eliminate or degrade them or otherwise decrease their usefulness to future generations, and implies using non-renewable natural resources at a rate slow enough to ensure a high probability of an orderly societal transition to new alternatives.

The notion of sustainable development places clear emphasis on intergenerational equity – fair treatment between present and future (unborn) generations. In other words, future generations should not be worse off than present generations and any development should be consistent with such long-term responsibilities. Sustainable development involves an increased emphasis on the value of the natural, built and cultural environments and the recognition that all three are interlinked. The determination of equivalent value is a vital ingredient in the economic appraisal of development projects.

When evaluating building projects in terms of their environmental impact, consideration needs to be given to all the costs and benefits that flow from the decision over the life of the project. This immediately raises the question of how to equate costs and benefits that occur in different time periods. Discounting is a technique that has been developed to adjust for the effects of time by recognising a preference for events that occur in the near future over events that occur in the far future.

The rate of discount used is critical to the evaluation process and the validity of the outcome. There is a clear connection between the environmental crisis, economic development, the need for such development to be sustainable and the role that discounting plays in the determination of equivalent value. Much, therefore, relies on the integrity of the discounting process, and its further understanding will significantly contribute to knowledge in the field.

'A discussion of the arguments for discounting raises several problems. The literature is anything but clear, and there exists little consensus on the subject. Different theories lead to different conclusions, and the positions are hard to compare as the assumptions and approaches differ considerably ... discounting is one of the most misunderstood concepts in economic analysis ... few topics in our discipline rival the social rate of discount as a subject exhibiting simultaneously a very considerable degree of knowledge and a very substantial level of ignorance.'
(Angelsen 1991, p. 12)

Past analyses of design solutions for building projects have concentrated on initial capital costs, often to the extent that the effects of subsequent operating costs are completely ignored. However, even in cases where a wider view of cost has been adopted, the discounting process has commonly disadvantaged future expenditure so heavily as to make performance after the short term irrelevant to the outcome, resulting in precedence being given to projects which display low capital and high operating costs. Thus design solutions that aim to avoid repetitive maintenance, reduce waste, save non-renewable energy resources or protect the environment through selection of better-quality materials and systems, usually having a higher capital cost, are often rejected on the basis of the discounting process.

The factors influencing the choice of discount rate are of concern because sustainable development is possible only if the equivalent value of future performance can be appropriately assessed. Although economic progress and environmental conservation are not mutually exclusive, there is normally a trade-off, and a balance therefore needs to be struck. Bias towards one or the other will directly affect the quality of life of future generations. Discounting must facilitate the equitable assessment of both present and future events. Doubt exists over whether this is currently the case.

Yet on the whole economists have acted as if there was no controversy and no broad constituency of concerns; they have routinely and mechanically applied the discounting technique in arenas of human activity very far removed from the financial markets in which they have their origin. There is great institutional convenience in the uniform adoption of a standard form of appraisal; it seems to offer consistency, even if only in the form of consistent error. Compromises have been sought that modify discounting incrementally so that its long-term effects are less severe. But most agree that discounting is so firmly entrenched that it cannot readily be displaced, even if such action were considered appropriate.

The discounting debate has progressed over the years from a rationale based on opportunity cost, through the psychology of time preference into wider fields of welfare and environmental economics. Discounting also raises fundamental issues about intergenerational equity and reflects contemporary sustainable development concerns.

THE CONCEPT OF VALUE

It is a defensible position to argue that the principal role of the building economist (or quantity surveyor), together with the entire design team, is to provide maximum value for money to the investor. But value is a difficult concept to define, particularly as it means different things to different people. The words cost and value are often interchanged in context, but it is important to be aware that cost is only a part, to some a minor part, of value perception.

Aristotle identified seven classes of value circa 350 BC that are still relevant to society. The classes of value can be summarised as being of an economic, moral, aesthetic, social, political, religious or judicial nature. Economic value may be seen as an objective consideration, measurable in terms of money, while the remainder are more subjective.

While not disputing Aristotle's classes of value, others have taken different approaches. For example, Arthur Mudge (a senior commentator on value management and a frequent writer on value perception and identification) concentrated on physical assets, isolating four components that may combine to form value:

- *Use value* This is the benefit attached to the function for which the item is designed.
- *Esteem value* This value component measures the attractiveness of the item.
- *Cost value* This represents the actual cost to produce and maintain the item over its period of possession.
- *Exchange value* This is the worth of the item as perceived by others who are interested in its acquisition.

Value includes subjective considerations that highlight the relationship between what someone wants and what he/she is willing to give up to get it. Value is thus relative and not an inherent feature of any object. Value commonly applies to assets, is measured by comparison with other assets of similar function, attractiveness, cost and/or exchange worth, and cannot be assessed in isolation.

The definition of value is often simplified by the building economist when undertaking comparative studies of buildings or their components. Maximum value is assumed to be found when a required service or function is attained and when the cost of providing that service or function is at a minimum. Value in this context can be measured objectively, but any solution found through such a procedure risks sub-optimisation. An increase above the required level of service or function for a small extra cost would usually be perceived by investors as 'better' value.

Value for money is therefore a balance between the subjective and objective considerations that the investor regards as important, the latter requiring some form of financial analysis. Nevertheless, any technique that is developed to measure costs must be viewed as contributing to but not encompassing the identification of value for money. In addition, subsequent monitoring and control of costs is necessary to ensure that the value objectives (or targets) are ultimately fulfilled.

Value for money can be computed by dividing a project's benefit by its total cost, and is therefore akin to a benefit-cost ratio. The benefit can be expressed as a value (index) score.

THE IMPORTANCE OF TOTAL COSTS

'It is widely accepted that to evaluate the costs of buildings on the basis of their initial costs alone is unsatisfactory. Some consideration must also be given to the costs-in-use aspects which will occur throughout the use of buildings. The proper consideration of the total costs of buildings is therefore likely to result in offering the client better value for money.' (Ashworth 1988, p. 17)

Commonly the measurement of costs is undertaken on a capital cost basis. Buildings are analysed in respect of their likely construction costs, to which might be added land purchase, professional fees, furnishings and cost escalation to the end of the construction period. Budgets similarly address initial costs, and the planning and control processes they foreshadow normally do not extend beyond hand-over.

The need to look further into a building's life than merely its design or construction period is a trivially obvious idea, in that all costs from an investment decision are relevant to that decision. The total cost approach takes into account both capital and operating costs so that more effective decisions can be made.

Society may be paying substantially more for its buildings over their whole life cycles than the theoretical (practical) optimum. Concerns of this type have led to continued interest in total costs, their impact and their quantification. There are numerous difficulties that have been advanced as being responsible for the lack of acceptance of the total cost approach by practitioners. Nevertheless most experts agree, conceptually at least, that the approach is clearly preferred to one that simply ignores the inevitable costs of ownership.

TERMINOLOGY

Terminology in common service includes life cycle cost, life-cost, recurrent cost, costs-in-use, operational cost, occupancy cost, running cost, ultimate cost and terotechnology.

The proliferation of terms has resulted from the progressive debate and redefinition of the total cost approach in a number of countries, principally the United Kingdom and the United States.

Life cycle cost is now the most popular term. The technique (life cycle costing) is best defined as the economic assessment of competing design alternatives, considering all significant costs of ownership over the economic life expressed in equivalent dollars. The technique takes account of both initial design and acquisition costs and subsequent running costs and is measured over the asset's 'effective' life, generally considered as its economic or operating life. Although many definitions exist, the majority exclude matters of revenue generated by the asset, other than taxation concessions and salvage value.

The term 'life cycle' has a well-established meaning in the field of biology. The image of a life cycle is one of development and renewal of an organism over time, so an analogy is used when employing the term to describe inanimate objects such as buildings. The phases of a building life cycle can be described as planning and design development, documentation of design, construction, commissioning, furnishing, operation, maintenance, rehabilitation and demolition. These phases are derivative of those first published by the British Government as being applicable to all assets.

During discussion of PA Stone's first paper on costs-in-use at a meeting of the Royal Society, Mr Winsten (fellow) said 'I was a little worried by the phrase "costs in use" which seems ...to imply only the cost of keeping something running, and not the cost of construction. Is there not an alternative similar short phrase? "Costs over time" might do.' (Stone 1960, p. 266)

Terotechnology comes from the Greek 'terein' meaning to look after, and is generally defined as a combination of management, financial, engineering and other practices applied to physical assets in pursuit of economic life cycle costs. It is a multi-disciplined approach to ensure optimum life cycle costs in the development and use of equipment and facilities and encompasses the management of assets from creation through to disposal or redeployment.

Life-cost is a term that has seemingly independent origins in Canada, Australia and New Zealand. As with life cycle costs, it is used to describe the sum of all expenditure relating to a building or other asset over a specified duration, although the nature of this duration lacks consensus. On face value this term appears less clumsy and more meaningful than previously described nomenclature, but to date it has received no widespread acceptance.

Other terminology either does not adequately describe the inclusion of capital costs, is too specialised or too vague. *Costs-in-use* is the most notable of these remaining terms, principally because it was the first to be adopted and widely discussed. Costs-in-use is the term applied to the summation of the operating costs of a building (such as fuel and power), the maintenance costs and the capital costs, and thus represents the true costs to the owner or occupier of the building as he/she uses it.

The temptation to dismiss life cycle costing techniques as being merely costs-in-use techniques with different terminology must be resisted. Life cycle costing is a total management tool for long-term ownership costs and a framework within which a formalised structure of cost planning can exist, and has evolved from intensive research and development activity in many sectors.

Life-cost or life cycle cost are considered the most appropriate descriptors, since it is essential that any approach aimed at measuring these costs must refer to a clearly identified and continuous period of time (or life). But 'life cycle' implies that the determination of life is based on a 'cradle to grave' approach, which in many cases forms an inappropriate time frame for the measurement of ownership commitment.

Life-cost studies shall be held to mean any activity using the total cost approach, and in particular may be considered as comprising life-cost planning and life-cost analysis. These terms will be discussed more fully in Chapters 4 and 5.

Life-cost is the term used hereafter in this book. The choice of terminology at this stage is purely to facilitate consistency, although it is personally preferred. While some may question 'what's in a name?', it appears from history that the introduction of new terminology in this field is directly related to revitalised interest in its application.

SUSTAINABLE DEVELOPMENT

The philosophy of sustainable development borrows freely from the science of environmental economics in several major respects. A basic component of environmental economics concerns the way in which economics and the environment interact. Fundamental to an understanding of sustainable development is the fact that the economy is not separate from the environment. There is an interdependence because the way humans manage the economy affects the environment and environmental quality, in turn, affects the performance of the economy.

The risks of treating economic management and environmental quality as if they are separate, non-interacting elements have now become apparent. The world could not have continued to use chlorofluorocarbons (CFCs) indiscriminately. That use was, and still is, adversely affecting the planet's natural ozone layer. Furthermore, damage to the ozone layer affects human health and economic productivity. Few would argue now that we can perpetually postpone taking action to contain the emission of greenhouse gases (GHGs). Our use of fossil fuels is driven by the goals of economic change, and that process will affect global climate. In turn, global warming and sea-level rise will affect the performance of economies.

Definitions of sustainable development abound, since what constitutes development for one person may not be development or progress for another. Development is essentially a value word: it embodies personal aspirations, ideals and concepts of what constitutes a benefit for society. The most popular definition is the one given in the Brundtland Report.

Sustainable development is defined as 'development that meets the needs of the present without compromising the ability of future generations to meet their own needs'. (Brundtland Report 1987)

This definition is about the present generation's stewardship of resources. This means that for an economic activity to be sustainable it must neither degrade or deplete natural resources nor have serious impacts on the global environment inherited by future generations. For example, if greenhouse gases build up, ozone is depleted, soil is degraded, natural resources are exhausted and water and air are polluted, the present generation clearly has prejudiced the ability of future generations to support themselves.

Sustainable development in practice represents a balance (or compromise) between economic progress and environmental conservation in much the same way as value for money on construction projects is a balance between maximum functionality and minimum life-cost. The economy and the environment necessarily interact, and so it is not appropriate to focus on one and ignore the other.

Development is undeniably associated with construction and the built environment. Natural resources are consumed by the modification of land, the manufacture of materials and systems, the construction process, energy requirements and the waste products that result from operation, occupation and renewal. Building projects are a major contributor to both economic growth and environmental degradation and hence are intimately concerned with sustainable development concepts.

Having said that, there is probably no such thing as sustainable development for the general run of construction projects. But that is not to say that consideration of sustainability is a waste of effort. On the contrary, every project (new or existing) can be enhanced by consideration of whole-of-life methodologies, particularly during pre-design. While most projects will consume more resources than they create, projects which are closer to sustainable ideals will increasingly deliver benefits to their owners and users and to society as a whole. Therefore, if design can encompass assessment and decision-making processes that address sustainability goals, it is likely that over the long term the construction industry will be able to demonstrate a significant contribution to global resource efficiency.

FORECASTING FUTURE EVENTS

The forecasting of future events is normally an integral part of the decision-making process. It also can be the subject of considerable uncertainty and therefore requires cognisance of the level of risk exposure. Discounting and life-cost studies are clearly reliant on appropriate forecasts of future events being made.

'The value of the technique lies in the rigour introduced into the evaluation, the need to make all assumptions explicit and the separation of real and chance effects.' (Stone 1975, p. 70)

The decision-maker is faced with a fundamental choice. Either the uncertainty of the future can be ignored by dealing with only those matters which are known, or estimating tools can be used to make predictions. It is frequently accepted in the literature that it is preferable to plan even when the accuracy of the plan cannot be guaranteed, for otherwise decisions are made in isolation from the environment in which they will ultimately be judged.

A life-cost 'plan' is an expectation of future design performance and hence forecasts events that are inherently uncertain. Even capital costs display a level of uncertainty, but when coupled with costs over a long period of time that themselves are a function of economic factors, obsolescence and operating performance, the entire process may appear an exercise in futility. But the life-cost plan should be seen as a set of targets against which future events can be compared. While the quality of the original prediction is important, a far greater benefit can be realised if areas of poor performance or areas in which additional savings can be made are identified.

Comparative life-cost studies involve further aspects of uncertainty, notably the selection of the discount rate and the period of financial interest of the owner. But even these areas of uncertainty can be diminished to some extent since they always apply to a range of solutions and hence can have a corresponding effect on estimated costs and benefits.

Yet it is the adoption of risk analysis techniques that enables the uncertainty of future events to be properly assessed. If all the outcomes are favourable then what is at risk is the level of the benefit, not the possibility of a loss.

However, investment decisions are not always so simple, and therefore a number of sophisticated risk analysis techniques are available to quantify the impact of various future scenarios. Risk analysis is a field in itself, and it is not intended to explore it further at this time. Yet its application to the determination of equivalent value is undeniable.

There are various techniques available to assist in the forecasting of future events. These include extrapolation based on time series or regression, probability distributions, simulations and the like. The choice of technique is a function of the type of forecast being made and the data that are available. But, in general, forecasts are made on the basis of historical evidence and hence there is an implicit assumption that the future will be generally consistent.

Forecasting models are abstractions from reality. Models focus on a few characteristics that best represent the real system in a way that is simple enough to understand and manipulate yet similar enough to reality to permit satisfactory results whenever the model is used in decision-making. The test of a model's performance is whether or not it can provide reliable predictions when viewed later using hindsight. Since models must be continually revised, forecasting implies a learning process in which the original model is formulated and the parameters are estimated from empirical data. If the predicted result varies widely from the actual outcome, a critical appraisal must be made to ascertain whether the underlying assumptions have been seriously violated. If the assumptions are fallacious, the model should be re-simulated to determine whether the ex post forecast is accurate given the proper set of assumptions. On the other hand, if the predictions are still at variance with actual results, the model must be either recast or replaced. Thus forecasting is a continual process of technique selection, model construction and forecast evaluation.

Most forecasting techniques will inevitably fail to predict catastrophic events but instead will focus on a range of outcomes that may be reasonably concluded from history. This results in the identification of best, worst and most likely outcomes. If a decision is insensitive to this range then it has a lower level of risk than if it were to change easily. Forecasting adds information to the decision-making process that is a vital part of proper analysis. It is clearly an integral part of a total cost approach.

REVIEW QUESTIONS

1.1   Is society paying more for its buildings due to our past failure to consider total costs?

1.2   Why has life-costing been relatively successful in the United States but not in most other countries?

1.3   Are there differences between the terms 'life-cost', 'life cycle cost' and 'cost-in-use'?

1.4   What benefits do you consider would flow from widespread consideration of total costs on construction projects?

1.5   When was the total cost approach first seriously proposed for building evaluation?

1.6   What are the four types of value developed by Arthur Mudge?

1.7   Can life-costs include consideration of revenue?

1.8   What is the difference between cost and value?

1.9   Why is sustainable development of interest to the construction industry?

1.10   What is the current motivation for interest in life-cost studies?

TUTORIAL EXERCISE

The graph on the next page illustrates the movement in world oil prices from 1970 to 2002, and the table following lists some specific data from 1986 to 2002. (World oil prices can be researched from the US Energy Information Administration: http://www.eia. doe.gov.)

Construct a prediction for the world oil price for 2005 (in US$/barrel) and provide some discussion about the confidence of your prediction. You can use Microsoft Excel to assist if you wish.

World Oil Prices (US$/barrel)

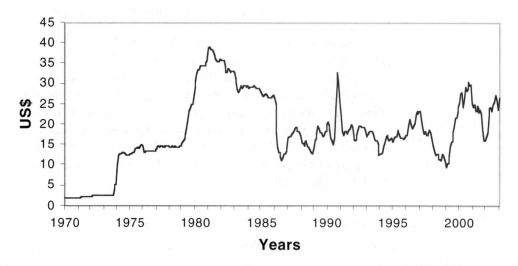

World Oil Prices 1986-2002 (June Figures)

Year	Price (2002 US$/barrel)
1986	12.25
1987	18.71
1988	15.50
1989	18.27
1990	15.15
1991	17.78
1992	19.83
1993	17.70
1994	17.04
1995	17.44
1996	19.03
1997	17.38
1998	11.67
1999	15.91
2000	28.91
2001	23.95
2002	23.45

HINT
In order to forecast from a series, a regression analysis (line of best fit) is commonly used. The regression equation enables values along the regression line to be calculated based on values of X (years), Y-intercept (US$/barrel) and slope of the line (% change).

KEY POINTS

A PowerPoint presentation dealing with the topics discussed in this chapter can be downloaded from the publisher's website (see the publisher's details at the beginning of the book for the address). Some key points are shown below.

Total Cost Approach

* ❃ Commonly the measurement of costs is undertaken on a capital cost basis
* ❃ The total cost approach takes into account both capital and operating costs so that more effective decisions can be made
* ❃ Concern that evaluation based on capital costs alone may lead to society paying more for its buildings than the theoretical optimum
* ❃ Still not widespread use in many countries

Terminology

* ❃ Life-cost (preferred)
* ❃ Life cycle cost (most common UK and US)
* ❃ Costs-in-use (original term)
* ❃ Operational cost, running cost, total cost analysis
* ❃ Occupancy cost, functional-use cost
* ❃ Ultimate cost
* ❃ Terotechnology , whole-of-life cost
* ❃ Recurrent cost

First Generation Technique

❋ Although the idea of discounted cash flow analysis has been around for over a century, the total cost approach for buildings was first seriously proposed in 1960 (Stone)

❋ The reason for its interest was that in the UK there was concern at this time over the maintenance burden of the country's aging stock of buildings

❋ Maintenance was therefore behind the first generation use of the technique

Second Generation Technique

❋ The technique was often discussed and advocated but never became common practice

❋ It took the world oil crisis of the 1970s to revitalise interest in the technique, particularly as there was great concern that the enormous increases in fuel prices could not be afforded in the future

❋ Energy conservation was therefore behind the second generation use of the technique

Third Generation Technique

❋ While the technique became popular in the US, mainly due to legislation, most other countries lost interest again as world oil prices stabilised

❋ In the late 1980s the need to use non-renewable resources in a sustainable manner renewed interest in the technique as a means of measuring building performance

❋ Environmental protection is driving the third (current) generation use of the technique

Life-Cost Studies

❊ Total asset management must be seen as the ultimate bounds of the life-cost technique

❊ This involves project initiation processes, investment analysis, project management and facility management

❊ Life-cost studies are vital to total asset management and will become common practice as governments become more concerned over resource usage and sustainability

ANSWERS TO REVIEW QUESTIONS

1.1   Given that operating costs can significantly outweigh capital costs over a building's economic life, and given that operating costs are generally ignored from design decisions, the chance of sub-optimal decisions is very high.

1.2   Life-cost studies are required by legislation on all large public sector projects throughout the US, but other countries have generally avoided legislative intervention.

1.3   There is little difference between these terms, since they refer to the total costs of an asset from concept to demolition (cradle to grave), although life-cost is now the preferred term in many parts of the world.

1.4   There are many project-related benefits, but overall widespread use would encourage a better balance of capital and operating expenditure, which would help to achieve sustainable development goals.

1.5   P.A. Stone first introduced the technique 'costs-in-use' in a paper delivered at a meeting of the Royal Society in London in 1960.

1.6   Mudge's value types comprise use value (functionality), esteem value (attractiveness), cost value (production) and exchange value (perceived worth by others).

1.7   Some authors advocate that the life-costing technique includes revenue (including but not restricted to differential revenue), but in reality revenue is limited to special cash flows like residual value, tax depreciation, etc. and is therefore more like a 'negative' cost.

1.8 The cost of an asset is the expenditure involved in its production and manufacture, whereas its value is what someone in the marketplace is willing to pay in order to acquire it.

1.9 Construction of buildings and other structures is the largest development activity on the planet, and it is vital that the industry understand and implement sustainable practices, although government intervention will likely force this outcome in the near future.

1.10 The environmental movement and the desire to use resources wisely and in a sustainable fashion (as much as possible) are behind the current revitalisation of life-cost studies.

REFERENCES

Angelsen, A (1991), *Cost-Benefit Analysis, Discounting and the Environmental Critique: Overloading of the Discount Rate?*, Chr. Michelsen Institute, Department of Social Science and Development, Norway.

Ashworth, A (1988), 'Making life cycle costing work', *Chartered Quantity Surveyor*, April, pp. 17–18

Stone, PA (1960), 'The economics of building design', *Journal of the Royal Society*, vol. 123, part 3, pp. 237–273.

Stone PA (1975), 'Building design evaluation and long term economics', *Industrialization Forum*, vol. 6, no. 3–4, pp. 63–70.

World Commission on Environment and Development (1987), *Our Common Future* (Brundtland Report), Oxford University Press.

CHAPTER 2

THE TIME VALUE OF MONEY

LEARNING OBJECTIVES
In this chapter you will learn about the principles of time valuation or, in other words, how the value of money changes over time. The concept of discounting is introduced here. By the end of this chapter you should be able to:
- describe the key exchange rates used in time value calculations,
- be conversant with the main formulae applied to these calculations,
- understand the purpose of discounting, and
- use discount rates to express money in different time periods.

COMPOUND INTEREST

'A large body of literature exists on life cycle techniques involving the principle of compound interest. The operation of compound interest is fundamental to all formulae used in life cycle costing and reflects the time value for money concept – namely that money has the capacity to multiply with time.' (Flanagan 1984, p. 199)

Compound interest is the result of earning interest on interest, as may be experienced by money deposited with a financial institution (bank, building society, etc.) where the interest received is reinvested. The higher the number of compounding periods per annum, the faster the rate at which funds will grow.

The formula for calculation of the accumulated sum of a principal continuously compounded is:

$$S = P(1 + i)^n \qquad \text{(Eqn 1)}$$

where: S = accumulated sum ($)
P = principal ($)
i = interest rate per period (%) divided by 100
n = number of compounding periods p.a.

The principal represents the value of the original investment in today's dollars, while the accumulated sum is an amount that will be received in the future when the investment is terminated and the funds withdrawn. The concept of compound interest forms the basis for determining the time value of money and underlies all discounting formulae.

But while compound interest results in the growth of available funds, inflation has the reverse effect through reduction of general purchasing power.

INFLATION

Inflation is an economic term that describes the rate of increase in the price of goods and services over time caused by the relative balance between the forces of supply and demand. It can be defined as the persistent tendency for the general level of pricing to rise. This is often referred to as general inflation, as distinct from the specific inflation (escalation) that affects only certain commodities.

Clearly 'money today' is not equivalent and cannot be compared with 'money tomorrow' unless an adjustment is made for inflation. Converting dollars into a value comparable to money today is known as expressing it in real terms. The inflation rate is thus a kind of exchange rate for determining the time value of money.

Specific inflation leads to the identification of differential price level changes (drift) relative to general inflationary trends for those items so affected. Such changes may cause positive or negative adjustment to general inflation in the short term, but it is likely that these adjustments will be minimal when measured over a longer time frame.

Energy costs are often highlighted as being susceptible to positive drift as limits to the world's natural resources are approached. Evidence of this was seen in the 1970s, causing a corresponding interest in the analysis of operating costs of buildings.

Inflation has a compounding effect. If the inflation rate was a constant 10% per annum, a given dollar amount would increase exponentially when measured over a series of consecutive years (i.e. 10% of an increasingly larger amount). Economists usually have no compunction about postulating models that imply an exponential growth to infinity.

The future cost, measured in inflated dollars, of either an inflation-linked or fixed payment or receipt expressed in today's dollars can be determined for any point in time using a derivative of Equation 1, with inflation acting as the exchange rate:

$$FC = PC(1 + f)^n$$ (Eqn 2)
(inflation-linked)

or $$FC = PC$$ (Eqn 3)
(fixed)

where: FC = future cost ($)
PC = present cost of payment or receipt ($)
f = inflation rate p.a. (%) divided by 100
n = number of years

'Just as it would be nonsensical to add £3 + $5 + DM7 to equal 15 in the same way adding £2 which arises now, with £5 that arises in 12 months' time, with £6 that arises in six years' time, to total £13, would be nonsensical. Money which arises at different points in time, just like money in different currencies, cannot be directly compared.' (Przybylski 1982, p. 454)

Future cost is used in preference to accumulated sum in this instance as the latter implies reinvestment of capital which is not necessarily appropriate to projects involving environmental and social factors. Future value and accumulated value are not considered suitable substitutes for either term.

Determination of the actual cost (or 'cheque book' amount) expected at some future time is useful, but it is clearly more useful to evaluate this cost in relation to monetary values that are understandable. Value is a relative concept and must be expressed in equivalent (constant) dollars so that comparison with other values at different points in time can occur. Although any point in time can serve as a datum for this purpose, it assists interpretation if the datum is 'today'. It therefore follows that whenever the term value is used it is understood to be in real terms.

Inflation results in an increase in the cost of goods and services over time, but the relationship between the value of these goods and services and others is theoretically unaffected. In reality there are many events that can occur to change our perception of the relative value of particular items. However, general inflation is not one of them and can hence be ignored when making comparisons. For example, the cost of painting a door may be $25.00 today. In ten years' time it may cost $65.00 to complete the same work, but its value may be unchanged.

As time passes, future costs that are fixed become comparatively less in worth than future costs that are inflation-linked. Yet the real value of any future cost can be found by reducing it by the general inflation rate:

$$PV = FC(1 + f)^{-n}$$ (Eqn 4)

where: PV = present value ($)
FC = future cost of single payment or receipt ($)
f = inflation rate p.a. (%) divided by 100
n = number of years

Note that raising an expression to the power of -n is equivalent to the reciprocal of the expression raised to the power of n. Thus the following is also true:

$$PV = \frac{FC}{(1 + f)^n}$$

Value may be considered to include only that portion of total value that relates to the cost to produce or maintain an item. This assumption has a practical benefit since it enables value to be objectively described in monetary terms. Alternatively, present value can be considered to also include subjective matters assigned a dollar value, but defining total value in this manner is often impractical.

THE DISCOUNTING PROCESS

Investment return, such as interest earned on funds deposited with a financial institution, may also be employed as an exchange rate between money today and money tomorrow. Consider the case in which an individual receives a sum of $1000 and invests it at a return of 10% per annum compounded, such as may be attained from a typical bank account. Ignoring taxation at this stage, the $1000 will grow to $1100 at the end of the first year, $1210 at the end of the second year, and so on. It is assumed that this return represents the individual's best use for the funds and is risk free.

In such a case the account holder should value $1100 in one years' time, $1210 in two year's time, etc. as equivalent to $1000 now (i.e. its present value). The reduction of future dollars to its equivalent in money today is known as discounting and leads to the determination of an 'equivalent' present value, herein called discounted present value. The resultant value is in real terms since the investment return includes a component for inflation. The formula for discounted present value is:

$$DPV = FC(1 + i)^{-n} \qquad \text{(Eqn 5)}$$

where: DPV = discounted present value ($)
FC = future cost of single payment or receipt ($)
i = investment return p.a. (%) divided by 100
n = number of years

Investment return is used in lieu of inflation as the exchange rate. This formula will discount a cost or revenue at a point in time given by n to its equivalent present value or cost. Often, however, costs are incurred or revenues are received on a regular annual basis, and it becomes tedious to determine the discounted present values for each amount over a long duration. In this case the annuity formula can be employed to simplify the calculation:

$$DPV = \frac{FC[1 - (1 + i)^{-n}]}{i}$$

(Eqn 6)

where: DPV = discounted present value ($)
FC = future cost of annual payment or receipt ($)
i = investment return p.a. (%) divided by 100
n = number of years

The discounting process creates a negative exponential trend in the equivalent value of future cost. Equations 5 and 6 both assume yearly payments, as this is normally accurate enough for most applications. However, if discounted present value amounts are required on a six-monthly, quarterly or even monthly compounding basis, then the investment return, i, is divided by the number of compounding periods per year and the number of years, n, is multiplied by same. This action would increase the rate of exponential decline.

It is necessary that some adjustment of future costs be made. Of the mechanisms available for this purpose, 'most practitioners and experts accept that the [discounted] present value method produces the fairest and most rational results for comparison of different cash flows sensibly'. (Bowen 1976, p. 6)

Discounting recognises that the time value of money is affected by the level of interest earned on available funds. It is a procedure for presenting all costs in one time frame so that a direct comparison can be made between one alternative and another. It is erroneous to ignore the timing of future costs and merely add them to initial cost.

Discounting is thus relevant to the appraisal of costs over time. Time has been described in an economic sense as the fourth dimension of buildings. Methods of discounting future expenditures have a bearing on the basic assessment of future costs, and their significance in decision-making causes them to have major economic consequences in practice, whether for the individual, an enterprise or the whole economy.

DISCOUNT RATE

Equations 4 and 5 can be conveniently resolved and simplified to remove future cost from the calculation of discounted present value and hence negate the need to deal with inflated dollars. This involves the combination of both inflation and investment return exchange rates, illustrating the manner in which money is normally 'working' to accrue interest or dividends that help to offset the devaluing effects of inflation. A new exchange rate is thus created, herein called the discount rate, which is merely a 'real' rate of return expressed as a factor. It can be calculated as follows:

$$d = \frac{1+i}{1+f} - 1 \qquad \text{(Eqn 7)}$$

where: d = discount rate p.a. (factor)
      i = investment return p.a. (%) divided by 100
      f = inflation rate p.a. (%) divided by 100

A suitable approximation of this rate is d = i - f. The discount rate may vary in magnitude depending on the investment return, taxation liability (where considered) and application.

The previous formulae (Equations 5 and 6) for discounted present value therefore take on the following forms:

$$DPV = PV(1 + d)^{-n} \qquad \text{(Eqn 8)}$$

where: DPV = discounted present value ($)
      PV = present value of single payment or receipt ($)
      d = discount rate p.a. (factor)
      n = number of years

and
$$DPV = \frac{PV[1 - (1 + d)^{-n}]}{d} \qquad \text{(Eqn 9)}$$

where: DPV = discounted present value ($)
      PV = present value of annual payment or receipt ($)
      d = discount rate p.a. (factor)
      n = number of years

Discounted present value can be expressed as an annual equivalent (present value) by making PV the subject of Equation 9, as follows:

$$AE = \frac{DPV \cdot d}{1 - (1 + d)^{-n}} \qquad \text{(Eqn 10)}$$

where: AE = present value annual equivalent ($)
      DPV = discounted present value ($)
      d = discount rate p.a. (factor)
      n = number of years

'The introduction of the rate of discount into cost predictions allows both for the inevitable cost of borrowing money and of not investing money elsewhere, and at the same time provides a means for "discounting" the future.' (Stone 1973, p. 50)

It is important to be aware that the discount rate is correctly applied only when present value is the subject of the discounting process. If future costs are used in lieu then inflation is explicitly stated and the discounting process must use a nominal investment return. While the use of a discount rate simplifies calculation of the time value of money, two reservations need to be made very clear:

- Future costs that are expected to remain fixed over time, that is will not inflate, need to be adjusted by a discount rate based on zero inflation (i.e. f = 0) to correctly reflect their declining value.

- It may be necessary to apply differential price level changes to present values which are expected to rise or fall faster than the general inflation rate.

Both of these reservations may lead to a number of alternate discount rates used to measure the value of goods and services, which will complicate use of the technique.

'Cost-in-use calculations are made in real terms. Real interest rates are used, that is the expected rate after allowance for inflation. Similarly, real prices are used, that is the prices expected for the period to be covered by the calculation, discounted for inflationary rises but with allowance for expected changes in real prices.' (Stone 1983, p. 204)

While both discount rates (real) and investment returns (nominal) are commonly used in practice, care must be taken that the two approaches are not inadvertently mixed. The use of real values and real discount rates is recommended.

The determination of present value is easier and more understandable than future cost and is thereby preferred as the subject of the discounting process. It must then follow that a real rate of discount be employed, but subject to the above reservations.

The prediction of both investment return and inflation is fundamental to the calculation of the time value of money. The discount rate is advocated as the appropriate exchange rate between money today and money tomorrow for all investment projects. The formulation of a discount rate for a particular project, however, may be affected by additional matters to those described above, such as the ratio of equity to borrowed capital, taxation rates, profit expectations and risk contingencies, all of which shall be investigated in the next chapter.

REVIEW QUESTIONS

2.1   What is the difference between general inflation and specific inflation?

2.2   How can differential price level changes be measured to assist prediction?

2.3   Why is inflation largely irrelevant in the determination of present value?

2.4   Can present value be extended to include a monetary allowance for subjective matters like appearance and functionality?

2.5   What is the difference between nominal and real discount rates?

2.6   What is the purpose and application of present value annual equivalents?

2.7   Why do we choose to work with present value rather than future cost?

2.8   Are discounted present values real and can they be added?

2.9   Why is it incorrect to discount present day costs with a nominal discount rate?

2.10   What are PV tables?

TUTORIAL EXERCISE

Construct a PV table for discount rates from 1% to 5% inclusive covering a period of 30 years. The values should be expressed in terms of the present value of $1 in each of these years. Prepare a graph showing the effect of discounting at each of the above discount rates.

> **HINT**
> A PV table is a list of discounting factors (expressed as DPV) for various discount rates based on a PV of $1.

KEY POINTS

A PowerPoint presentation dealing with the topics discussed in this chapter can be downloaded from the publisher's website (see the publisher's details at the beginning of the book for the address). Some key points are shown below.

Compound Interest

* ❋ Compound interest is the result of earning interest on interest
* ❋ The concept of compound interest forms the generally accepted basis for determining the time value of money and underpins all life–cost formulae
* ❋ Money grows in value with time because of investment potential
* ❋ The rate of growth follows an exponential curve

Inflation

❈ Inflation has the opposite effect to interest

❈ It is an economic term that describes the rate of increase in the price of goods and services over time

❈ 'Money today' is not equivalent and cannot be compared to 'money tomorrow' unless an adjustment is made for inflation

❈ Converting future dollars into its equivalent today is called expressing it in real terms

Future Cost

❈ Future cost is the inflated cost of today's value

❈ It is the 'cheque book' amount needed to pay for something in the future

❈ Future cost cannot be accumulated across time periods as costs in different years first need to be brought to a common base

❈ Present cost and future cost are normally different unless the future cost is fixed and not affected by inflation

Present Value

❈ Inflation results in an increase in the cost of goods and services over time, but the relationship between the value of these goods and services and others is theoretically unaffected

❈ For inflation-linked expenditure, present value and present cost are the same

❈ For fixed expenditure, present value is less than present cost

Discounting

* ❋ Investment return, such as interest from a bank, may be used to compare the value of money now and in the future

* ❋ Ignoring taxation issues, $1000 today equals $1100 in one year's time, which equals $1210 in two years' time and so on (assuming 10% interest)

* ❋ Future cost can be discounted by the investment return because this new 'value' is all that is needed to pay for the intended future cost

Discounted Present Value

* ❋ The theoretical value that needs to be set aside today to pay for a future cost is called discounted present value

* ❋ Discounting recognises that the time value of money is affected by the level of interest earned on available funds

* ❋ Since investment return and inflation are both involved and working in opposite directions, they can be combined into a single exchange rate

Discount Rate

* ❋ The combination of investment return and inflation is called the discount rate

* ❋ It acknowledges that money is normally 'working' to accrue interest or dividends that help to offset the otherwise devaluing effects of general inflation

* ❋ With the use of real discount rates, present values can be converted into discounted present values for use in economic evaluations

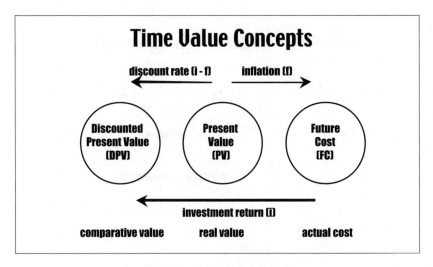

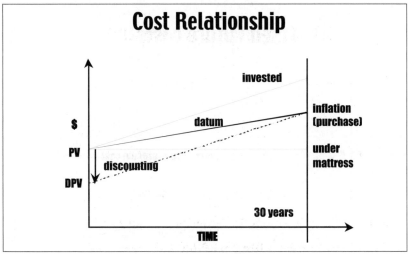

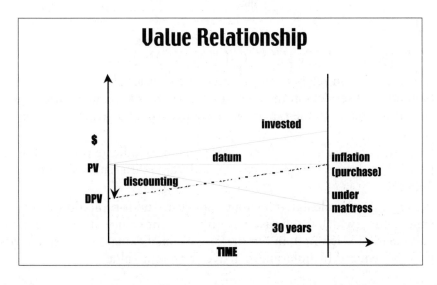

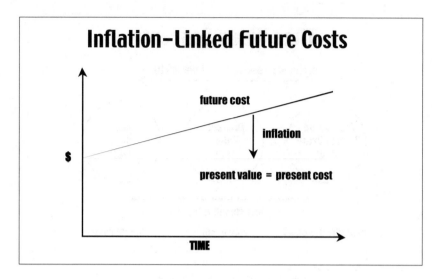

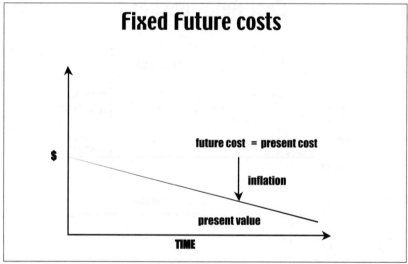

ANSWERS TO REVIEW QUESTIONS

2.1 General inflation refers to the average rate of price change for a range of goods and services in the community, whereas specific inflation (or escalation) refers to the rate of price change for a particular good or service.

2.2 Past prices can be analysed and extrapolated to indicate possible future trends, though over long forecast periods large differences should be unlikely.

2.3 Present value is calculated from present cost, and for inflation-linked expenditure (which is the vast majority) present value equals present cost and inflation is ignored, but in the case of fixed expenditure present cost must be reduced by inflation to arrive at present value.

2.4   Yes, subjective matters can have theoretical monetary values placed on them and used in the normal way, but in practice this is rare as the valuation of intangibles can be quite difficult to assess and open to criticism.

2.5   A nominal discount rate is simply the investment return, whereas real discount rates are investment return less inflation, with the latter being preferred and applied in real cash flows (present value) to calculate discounted present value.

2.6   These are used in comparisons between alternatives with unequal lives or study periods so that the evaluation can proceed, and they are also useful as a way of expressing the average annual comparative value of alternatives in today's dollars.

2.7   Present value is easier to use because inflation is treated implicitly in the discount rate rather than explicitly in the determination of future cost, and subsequent risk analysis is simplified because the discount rate combines various economic forecasts and can be tested holistically.

2.8   Yes, discounted present values are real in the sense that they are in today's dollars and can be added, but no, they are not real in the sense that they are a theoretical value used only for comparison.

2.9   Since present day costs do not include any allowance for inflation, discounting them with a rate that does include inflation gives erroneous results.

2.10   PV tables are discount rate factors that give equivalent results to the formula $(1 + d)^{-n}$ based on a present value of $1.

REFERENCES

Bowen, B (1976), 'Applied life cycle costing' in proceedings of Canadian Institute of Quantity Surveyors Seminar, Ottawa, pp. 1–35.

Flanagan, R (1984), 'Life cycle costing: A means for evaluating quality' in *Quality and Profit in Building Design*, Brandon PS and Powell JA (eds.), E&FN Spon, pp.195–207.

Przybylski, EA (1982), 'Effect of inflation and rate of interest on cost-in-use calculations' in *Building Cost Techniques: New Directions*, Brandon PS (ed.), E&FN Spon, pp. 454–461.

Stone, PA (1973), 'The application of design evaluation techniques' in *Value in Building*, Hutton GH & Devonald ADG (eds.), Applied Science Publishers.

Stone, PA (1983), *Building Economy* (Third Edition), Pergamon Press.

CHAPTER 3

DISCOUNTED CASH FLOW ANALYSIS

LEARNING OBJECTIVES
In this chapter you will learn about the technique of discounted cash flow analysis and how it is applied to make comparative judgements about projects. The impact of taxation and its consideration in such studies is also discussed. By the end of this chapter you should be able to:
- prepare a discounted cash flow,
- understand how taxation is incorporated in the cash flow,
- describe the different approaches to discounting, and
- calculate and use selection criteria to make investment decisions.

CASH INFLOWS AND OUTFLOWS

Discounted cash flow (DCF) analysis is a technique for assessing the return on capital employed in an investment project over its economic life, with a view to prioritising alternative courses of action that exceed established profitability thresholds. Investment projects typically span more than one year and hence involve future costs and benefits. Since payments or receipts arising at different times have different worth per unit, their values must be expressed in equivalent dollars before they are comparable. DCF is extensively, although not exclusively, used for this purpose and may be evidenced in a range of economic appraisal techniques in practice.

The term cash flow normally represents transactions involving liquid assets, including those that arise from taxation liability, generated by a particular investment project. Cash inflows relate to benefits that are received and which can theoretically be deposited in a bank account, while cash outflows are expenses incurred in the course of obtaining those benefits. The net cash flow is simply the difference between the cash inflows and outflows and is essentially a function of cost, investment requirement and tax structure.

Economic life is defined as the period of time during which the investment is the least-cost alternative for satisfying a particular objective.

The discounted cash flow technique focuses on the overall cost consequences of an investment, considering the amount and timing of cash inflows and outflows and envisaged rates of return. The underlying principle is to determine the value of future cash flows generated by an investment opportunity over its economic life. This can preferably be done by applying an appropriate discount rate to reduce the present value of future costs to a value that reflects its investment potential today (i.e. discounted present value).

The purpose of discounting therefore is to enable decisions to be made regarding alternatives that exhibit different patterns of cash inflow and outflow. Table 3.1 illustrates this purpose through the comparison of three hypothetical investment opportunities. Each investment comprises a total of $300 000 over five years. The only difference between the investments is the timing of the cash flows. The cash flows can be taken as net benefits or as net costs since the discounting process operates correctly under both conditions. A 10% discount rate has been assumed.

Table 3.1 The Discounting Philosophy

Year	Investment A		Investment B		Investment C	
	PV ($)	DPV ($)	PV ($)	DPV ($)	PV ($)	DPV ($)
0	50 000	50 000	300 000	300 000	–	–
1	50 000	45 455	–	–	–	–
2	50 000	41 322	–	–	–	–
3	50 000	37 566	–	–	–	–
4	50 000	34 151	–	–	–	–
5	50 000	31 046	–	–	300 000	186 276
Total:	300 000	239 540	300 000	300 000	300 000	186 276

Discounting is used to select the best alternative. If the cash flows are considered as benefits, then Investment B would be preferred because it generates the highest benefit ($300 000). If the cash flows are considered as costs, then Investment C would be preferred because it generates the lowest cost ($186 276). Discounting is a mechanism that assists decision-making particularly when the cash flows are more complex and the optimal selection less obvious.

Investments that exhibit cash inflows (benefits) earlier in their life are preferable to investments where the cash inflows are delayed. Similarly, investments that exhibit cash outflows (costs) later in their life are preferable to investments where the cash outflows are accelerated. In other words, costs are best delayed for as long as possible and benefits are best received as early as possible. Discounting therefore supports a notion of time preference and favours situations where control and possession of money is maximised.

CAPITAL BUDGETING TECHNIQUES

Discounted cash flow analysis forms the basis for all capital budgeting techniques that take account of the changing value of money over time. Most experts agree that discounting is fundamental to the correct evaluation of projects involving differential timing in the payment and receipt of cash.

Accounting rate of return and simple payback methods, which do not consider changing time value,'are quite inadequate substitutes for the discounted cash flow methods, and in fact are quite likely to produce results that would cause incorrect investment decisions to be made if they were to be used in isolation'. (Peirson and Bird 1981, p. 110)

There are numerous applications of discounted cash flow analysis used in capital budgeting. The two most common applications are net present value (NPV) and internal rate of return (IRR). Both rely on the existence of costs and benefits and lead to the identification and ranking of projects that are financially acceptable.

Net present value is the sum of the discounted present values of all future cash inflows and outflows. If the net present value is positive at the end of the economic life, then the investment will produce a profit with reference to the discount rate selected. If the net present value is negative, then the investment will produce a loss. Mutually exclusive projects should be selected based on the magnitude of the net present value, provided it is positive.

The internal rate of return of an investment is the discount rate that, if chosen, will lead to a net present value of zero. It is commonly established by iteration or trial and error using a financial calculator or computer, or by graphical means. The primary purpose of IRR analysis is simply to reflect the relative productiveness of the capital being committed to the project under consideration, not to provide an absolute measure of profitability.

If this rate offers an acceptable return to the investor then the investment proposal presumably will be approved. If this rate falls below the investor's minimum acceptable return, the proposal presumably will be rejected. If several alternatives are available, those showing the highest percentage returns will be given top priority. Depending on the cash flow pattern, some projects may exhibit multiple internal rates of return while others may have no rate at all.

Table 3.2 is an example of a discounted cash flow analysis for a hypothetical investment project having an expected economic life of 10 years. The discount rate, expressed as a percentage, is 10% per annum. The net present value for this example is $57 992. The internal rate of return is calculated as 20.63% and represents the discount rate that, if used, would lead to a net present value in year 10 of zero.

Table 3.2 Example of DCF Analysis

Year	Cost (PV)	Benefit (PV)	Net Benefit (PV)	Discounted Net Benefit (DPV)	Cumulative Total DPV)
0	120 000	0	− 120 000	120 000	− 120 000
1	15 000	42 000	27 000	24 545	− 95 455
2	15 000	42 000	27 000	22 314	− 73 140
3	15 000	47 000	32 000	24 042	− 49 098
4	15 000	52 000	37 000	25 271	− 23 827
5	15 000	52 000	37 000	22 974	− 853
6	22 000	47 000	25 000	14 112	13 259
7	22 000	47 000	25 000	12 829	26 088
8	22 000	47 000	25 000	11 663	37 751
9	22 000	47 000	25 000	10 602	48 353
10	22 000	47 000	25 000	9 639	57 992

Net Present Value (NPV) = 57992

The most important factor in the calculation of net present value or internal rate of return is the value assigned to the various cash inflows and outflows used. They represent the 'expected' conditions that will occur, but should these conditions change the profitability of the investment can be significantly affected. For this reason, calculations should be repeated for cash flows that represent the 'worst possible' and 'best possible' conditions that could conceivably occur. The resulting variance will provide insight into the sensitivity of the investment and hence its risk. The larger the variance, the higher should be the return for the 'expected' conditions.

Figure 3.1 illustrates the simple relationship between internal rate of return and net present value. The project can be seen to be acceptable at any return that is less than 20.63%. Higher expectations would cause the project to be rejected.

Figure 3.1 Graphing the NPV and IRR Relationship

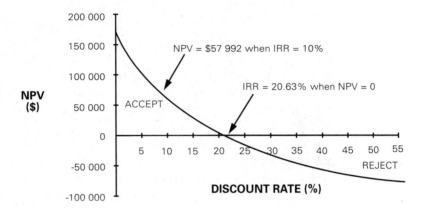

Other techniques for capital budgeting based on discounted cash flow analysis include the net-benefit (NB) method and the net-saving (NS) method, both of which look at the residue of benefits over costs (or vice versa) and discount this net amount over time. Two additional ratios may also be used to indicate economic performance, namely the benefit-to-cost ratio (BCR) and the savings-to-investment ratio (SIR). Furthermore, an approach called discounted payback (DPB) can be used to calculate the time needed before the project becomes economic, and this is compared to the useful life of the project. The project is considered viable if the DPB period is less than the useful life, provided no large costs occur after the DPB period has been reached.

COST-BENEFIT ANALYSIS

Cost-benefit analysis (CBA) is the practical embodiment of discounted cash flow analysis. There are two types of cost-benefit analysis: economic and social. Economic analysis involves real cash flows that affect the investor. Social analysis involves real and theoretical cash flows that affect the overall welfare of society.

Social cost-benefit analysis uses the concept of collective utility to measure the effect of an investment project on the community. Intangibles and externalities are assessed in monetary terms and included in the cash flow forecasts, even though in many cases there is no market valuation available as a source for the estimates. Nevertheless the technique is useful in that it attempts to take both financial and welfare issues into account so that projects delivering the maximum benefit can be identified.

Because social cost-benefit analysis is directed at maximisation of collective utility, it is generally employed as an investment technique by government agencies. Furthermore, the amount of research and estimation that the technique demands means that it is feasible for use mainly on large and complex projects, typically those involving infrastructure such as transport, health care, water irrigation, energy supply and the like.

More frequently the technique is being called on to evaluate environmentally sensitive projects, and has come under attack from many conservationists because it can justify investments that cause damage to the environment provided that this damage is outweighed by an increase in capital wealth. The quantification of intangibles and externalities clearly has great importance to the outcome. In addition, distributional effects are often overlooked, as the acceptance criterion is concerned with overall utility improvement and hence does not distinguish whether this benefit is evenly shared. For this reason projects that are likely to involve an uneven distribution of benefits should have a separate study undertaken to expose the effects.

The discounting process is extended in social cost-benefit analysis to adjust intangibles and externalities along with the more conventional cash flows. Utility is actually being discounted. For example, the value of lives saved from the introduction of a new hospital diminishes exponentially as time progresses. This is intuitively unsettling to many.

TAXATION

Taxation should be considered when using capital budgeting techniques because it plays a significant role in a project's economic performance. Therefore in DCF analysis all cash flows should preferably be treated net after tax. This implies that the discount rate and the internal rate of return are also net of tax (tax payers, whether using their own capital and sacrificing interest, or borrowing, effectively reduce the nominal interest rate by the prescribed taxation rate).

'The effect of tax reliefs on the rate of return ... can be dramatic, and any attempt to appraise an investment without taking the tax effects into account would produce very misleading results.'
(Wright 1972, p. 54)

Nevertheless, taxation is often ignored in discounting calculations in practice. The main justification appears to be that while income generates tax, so too does it generate tax deductions. Businesses tend to aim to minimise taxable income by effective organisation and accounting of their affairs. More likely, however, taxation is ignored for reasons of simplicity.

Taxation concessions are typically divided into either (a) deductions and rebates from taxable income or (b) depreciation. Deductions can be claimed for all expenditure incurred in earning income, and often include cleaning, energy usage, rates and charges and maintenance costs for income-producing assets. Interest payments on monies borrowed may also be deductible if negative gearing is allowed. Acquisition and subsequent renovation or replacement, however, cannot normally be deducted from income in this way. Depreciation is typically applied to the acquisition cost of certain assets, generally identified as plant and loose equipment, at rates that reflect their expected obsolescence.

Taxation policy and regulation varies by country and, even within a country, can frequently change over time. The information presented here is a general view and should always be interpreted in context.

It must be remembered that although certain operating costs are tax deductible and assets may be depreciable, their effect is dependent on the financial position of the controlling owner, be it a sole trader, partnership or corporation. Taxation relief is available only on profit and not every firm makes a profit every year, although losses can usually be carried forward. Taxation matters need to be analysed in terms of all holdings within the investor's portfolio, not merely those pertaining to a particular project. Government agencies do not pay tax at all.

Depreciation calculations are usually based on either the Prime Cost (PC) or Diminishing Value (DV) methods. The depreciation allowance applicable to either method for any year can be derived using the following formulae:

$$D_{PC} = PV \cdot v \qquad \text{(Eqn 11)}$$

(maximum v^{-1} years)

$$D_{DV} = PV.(1-v)^{n-1} \qquad \text{(Eqn 12)}$$

(theoretically infinite)

where: D_{PC} = depreciation (prime cost method)
D_{DV} = depreciation (diminishing value method)
PV = present value of acquisition with some exclusions ($)
v = depreciation rate p.a. (%) divided by 100
n = number of years

The discount rate chosen to convert the present value of yearly depreciation charges into their equivalent discounted value must assume that inflation is zero. Asset depreciation is typically based on the original acquisition cost, and it is not permitted to link this cost with general inflation movement. Depreciation is therefore a fixed cost and will devalue faster than those that are inflation-linked.

Capital gains tax is often payable on the inflation-adjusted profit (after selling expenses) received on sale of the asset, assuming that a profit was indeed made. Other types of taxation exist, such as goods and services tax, import duties, payroll tax, excises and fringe benefits tax, but will not be discussed in detail here.

THE RATIONALE FOR DISCOUNTING

'The problem areas generally arise as such because potential users of the technique are misled by the idea that the life cycle cost outcome must stand by itself in absolute terms. A little thought will lead to the conclusion that this is not possible. The life cycle cost outcome is only useful as a comparative figure for the purpose of ranking one solution over another. The same thing can be said of all the economic evaluation techniques which employ discounted cash flow analysis.' (Robinson 1986, pp. 18–19)

Discounting processes result in the calculation of 'equivalent' values that are useful in comparing alternative solutions involving cash flows that span over more than one year. Discounted present value is an 'abstract' measure that negates the need to explicitly consider finance costs while simultaneously allowing for the timing of cash flows. We need to remind ourselves consistently that we are comparing the worth of alternative systems and not their costs.

The necessity to take account of the time value of money when comparing future costs and benefits is clear, but the mechanism employed to achieve this is not. Discounting is advocated generally in the literature as being the appropriate conversion tool, and indeed has widespread acceptance in appraising investment projects in practice. Yet some 'mystic' and criticism remains.

One argument against the discounting process is that it renders future costs and benefits insignificant when a lengthy time horizon is involved. The traditional discounting method can reduce future costs to such an extent that their ability to influence decisions is essentially lost. But others conversely believe that this might be a convenient feature since the relative weight given to consequences declines with their degree of certainty, and thus the greatest weight tends to be given to those costs that are the most certain.

Others have criticised discounting on social grounds. As far as the balance between the present and future generations is concerned, discounting gives more importance to the former. Discounting may therefore lead to decisions being made that are inadvertently biased against future generations.

Future costs are worth less than their face value now because they can be secured by investing a smaller sum today that will grow with interest to the sum required at a particular date in the future. The effect of discounting is to estimate the value of this fund which, if provided at the inception of the project, would finance not only the initial capital investment but have enough left over, together with the interest it would earn when reinvested, to pay for the recurrent expenses over the life of the building. Of course, the reality of having such a fund available from the outset is incredible. Discounting must be seen as a theoretical value adjustment only.

'It can be argued that the choice of discount rate is one of the more critical variables in the analysis.' (Ashworth 1988, p. 234)

The choice of discount rate, if not made with due consideration to investment return, inflation and taxation, can lead to significant misinterpretation of decisions. Use of high interest rates based on project selection criteria has had the effect of minimising the benefit of increased operating efficiency and has fuelled speculation that discounting may not be totally appropriate. Large discount rates are a fallacy in today's economic climate, and perhaps always were so. Discounting with pure interest rates or borrowing rates is invalid unless the costs being discounted are first inflated. Ignoring income tax liability on investment return for non-government organisations is unrealistic, even though some companies may, from time to time, show a small net profit or even a loss at the end of a financial year.

While discounting has drawbacks, the value of the technique perhaps lies in the rigour introduced into the evaluation, such as the need to make all assumptions explicit and to separate actual from possible effects. Discounting is one method for objectively depicting the time value of money, but there is still no universal measure of comparison between costs incurred now and those incurred in the future. Discounted values are practically meaningless unless compared to that of an alternative solution

Would you prefer to receive $100 now or in a year's time? Would you prefer to pay your bills as soon as they are received or later when they fall due?

Nevertheless, most people would prefer to receive revenue sooner rather than later and would prefer to postpone expenditure for as long as possible. This is sometimes referred to in the literature as the personal rate of time preference or pure time preference. In order to increase the investment return, initial revenue must be maximised and initial expenditure minimised. Discounting future costs with a positive discount rate has the effect of favouring initial receipts and payments over those that occur subsequently. Assets that possess low initial costs and defer operating costs into the future lead to higher returns for the investor.

While there is a rationale for discounting as a process, the selection of the rate of discount is the subject of much controversy. There two distinct philosophies for discount rate selection. These are time preference and capital productivity.

TIME PREFERENCE

The philosophy of time preference is based on the assertion that people are impatient and seek control over money, assets and other resources. Individuals prefer the present to the future, even where monetary gain is not involved. As society is no more than the sum of individuals, society also prefers the present to the future. This temporal preference translates into a positive discount rate. Present consumption is therefore preferred to past consumption. Further analysis of the time preference philosophy reveals that there are at least three distinct supporting reasons. These are:

- *Pure impatience* Impatience (or myopia) is an undeniable human trait that results in present events being more desirable than those occurring in the future.
- *Risk and uncertainty* A benefit or cost is valued less the more uncertain its occurrence. This can result from uncertainty about the presence of the individual at some future date (risk of death), uncertainty about the preferences of the individual even when his or her existence can be regarded as certain, and uncertainty about the availability of the benefit or existence of the cost.
- *Intergenerational equity* If future generations are confidently expected to be 'better off', then the value of future events can be given less weight compared to the value of present events.

These factors may combine or be subsumed in some way to determine what is commonly identified in the literature as the social time preference (STP) rate of discount. Although impatience may well exist and be relevant to the calculation of equivalent value, it is difficult to quantify other than to assume that it can be reasonably modelled by use of normal investment return.

CAPITAL PRODUCTIVITY

Capital productivity is the discounting philosophy used predominantly in practice.

The philosophy of capital productivity is based on the theoretical earning capacity of money over time. While this phenomenon is real, there is little consensus on the method of calculation of the discount rate. Most opinions, however, can be reasonably aligned with one of the following assertions:

- The discount rate represents the minimum acceptable rate of return for the project.
- The discount rate represents the opportunity cost sacrificed by not investing in an alternate project.
- The discount rate represents the weighted cost of capital applicable to the financing of the project.

A project's net present value is often calculated using the required rate of return as the appropriate discount rate. If the net present value is positive, the project will generate at least the required rate and is thus acceptable. Determination of the minimum acceptable rate of return (MARR) may be based on a variety of factors, but will typically reflect the interest cost of borrowed funds (preferably after tax) plus an allowance for profit and risk. Some discount 'more risky' projects more heavily in order to allow for risk.

Alternatively the discount rate can be based on opportunity cost, defined as meaning the real rate of return on the best available use of funds that can be devoted to the project. Public projects often use discount rates based on the social opportunity cost of capital (SOC), which is interpreted as displaced private spending. The SOC is the marginal productivity of available capital and is identified as the best prospect for a practical approach to discounting for project analysis by some public sector authorities.

While the minimum acceptable rate of return and opportunity approaches concerning the choice of discount rate are not greatly dissimilar, the third assertion distinguishes itself by not specifically involving profit or risk. The latter attempts to reflect the true time value of money based only on explicit or implicit investment return rather than desirable levels of profitability.

Investors must either borrow money to finance the project or sacrifice interest earned on their own money. Use of investment return for discounting implies that interest on present equity for the full cost of the investment is being forgone, interest on borrowed funds for the full cost of the investment is being charged, or more likely a combination of both equity and borrowed funds is involved. Ideally the discount rate should be based on a weighted mix of the interest rate sacrificed by use of equity and the interest rate payable by use of borrowed funds, representing the expected financing arrangements for the project.

The discount rate applied to the weighted cost of capital approach remains determined by Equation 7, given that:

$$i = i_1 p_1 + i_2 p_2 \qquad\qquad \text{(Eqn 13)}$$

where: i = investment return p.a. (%)

i_1 = equity rate p.a. (%)

i_2 = borrowing rate p.a. (%)

p_1 = proportion of equity financing (%) divided by 100

p_2 = $1 - p_1$

Investment return, in other words, is a weighted mix of equity and borrowing rates in the proportion expected in the project over its duration.

The relative balance between capital and operating costs (if applicable), in present value terms, is useful in assessing the proportion of equity that will apply to the project. Capital costs are often raised through borrowings and operating costs from equity reserves built up from revenue.

Furthermore, the selection of an appropriate discount rate to be used in the calculations will largely depend on the financial status of each individual investor. The ratio of debt to equity financing for a company has been used to determine the weighted cost of capital. Whichever capital productivity approach is employed for determining the discount rate, it can be proven that they:

- negate the need to include interest payable on borrowed funds in the cash flow, as this is reflected in the establishment of the discount rate,
- favour expenditure in the far future over expenditure in the near future by enabling the saved equity to theoretically earn interest,
- are not affected by whether particular cash flows are financed from equity or borrowings provided that the overall mix of capital sources is appropriate.

THE CHOICE OF DISCOUNT RATE

Most literature suggests that the discount rate is normally based on capital productivity. Using this approach the discount rate reflects, at least in part, the interest payable on borrowed money to finance the investment and/or the interest lost through use of accumulated equity. A weighted combination of both interest rates is therefore involved. The interest rates may be real or nominal, before or after tax, and may include or exclude profit expectations and risk contingencies.

Matters of a non-financial nature are not well represented by the compound interest approach. It can be argued, for example, that a holiday in Hawaii is preferred now to one later in the year (all other factors being equal) yet this preference appears unrelated to the growth in reinvested income. Similarly, risk levels will influence the expected estimates for individual cash flows, perhaps by raising all costs by 10% and lowering all benefits by 10%, but this outcome is quite different from that resulting from a 10% increase in the discount rate.

On the other hand, financial matters can be realistically modelled using a compound interest approach. Interest payable for monies borrowed or interest lost due to equity usage both result in a burden that is proportional to time. The source of financing to be used for a project is relatively unimportant.

Interest can be incorporated into the discount rate or dealt with explicitly as a cash inflow or outflow. It may be correct to ignore discounting altogether provided an interest component is included in the annual cash flows as an expense for every year of the project. However, the use of a discount rate reflecting the weighted mix of equity and borrowing gives an identical result. There is no doubt that the inclusion of interest in both the cash flows and the discount rate is double-counting and incorrect.

The discounted present value of $124 869 is common to all three financing scenarios.

Table 3.3 demonstrates that the net present value for three identical investment projects is exactly the same (i.e. $124 869) when discounted by the weighted cost of capital, even though different financing arrangements are used and the net cash flow ranges between $130 000 and $163 000. Future cash flows are clearly worth less than their present value due to the changing time value of money. Although any scenario may be assumed, it is recommended that capital costs are allowed for and discounted in the years in which they are incurred. A 10% discount rate is used for the example calculation.

This exercise clearly illustrates that selection based on net present value will be unaffected by the decision to account for interest either in the discount rate or in the actual cash flows. It is merely convenient to exclude the detailed calculation of finance costs (both actual and imputed) from the appraisal by choosing a discount rate based on expected equity and borrowing rates mixed in their corresponding proportions.

Table 3.3 Example of Different Financing Methods

Scenario 1: 100% Equity

Year	Cash Requirement	Equity Consumed (A)	Borrowing Liability (B)	Net Cash Flow (A + B)	DPV at 10%
0	100 000	100 000	nil	100 000	100 000
1	10 000	10 000	nil	10 000	9 091
2	10 000	10 000	nil	10 000	8 265
3	10 000	10 000	nil	10 000	7 513
Total:	130 000	130 000	nil	130 000	124 869

Scenario 2: 100% Borrowed Funds

Year	Cash Requirement	Equity Consumed (A)	Borrowing Liability (B)	Net Cash Flow (A + B)	DPV at 10%
0	100 000	nil	nil	nil	nil
1	10 000	nil	10 000	10 000	9 091
2	10 000	nil	11 000	11 000	9 091
3	10 000	nil	*142 000	142 000	106 687
Total:	130 000	nil	163 000	163 000	124 869

* Includes $130 000 capital loan repayment

Scenario 3: 50% Equity and 50% Borrowed Funds

Year	Cash Requirement	Equity Consumed (A)	Borrowing Liability (B)	Net Cash Flow (A + B)	DPV at 10%
0	100 000	50 000	nil	50 000	50 000
1	10 000	5 000	5 000	10 000	9 091
2	10 000	5 000	5 500	10 500	8 678
3	10 000	5 000	*71 000	76 000	57 100
Total:	130 000	$65 000	81 500	146 500	124 869

* Includes $65 000 capital loan repayment

If the discount rate additionally includes an allowance for profit and risk, the previous equivalence between different financing methods is lost. Using the examples given in Table 3.3, but with a discount rate of 20% (i.e. 10% profit and risk added), the resultant discounted present values are $121 064, $98 148 and $109 606 respectively even though the investments are still identical. This highlights the problem encountered with discounting when profit and risk are incorporated in the discount rate.

A discount rate limited to the real weighted cost of capital reflects the true time value of money. The profitability of a project and its associated risk, of vital concern to an investor, can be independently assessed. Profit and risk expectations do not behave in an exponential manner, which is a good reason why they should be eliminated from the discount rate. Given that the entire discounting process is based on compound interest, it is reasonable when deciding on the composition of the discount rate to include only those components that are compatible.

It has now been established that it makes no difference whether interest is included in the discount rate or dealt with as an annual cash flow. An important point when interest is treated as a cash flow, however, is that interest lost through the use of equity must not be forgotten. Explicit calculation of interest must be divided into interest 'theoretically' received and interest payable. This approach on all but the most simple of investments will be extremely complex and hence highly subject to error. Since interest is intended in part to offset the devaluing effects of inflation and is expressed in 'future cost' terms, the interest cash flows would need to be reduced by the expected inflation rate. Furthermore, testing the sensitivity of assumptions in isolation or combination will take on overwhelming proportions. For all these reasons it is desirable to allow the discount rate to incorporate interest transactions.

If interest was to be treated as a distinct cash flow then the entire discounting process might well be redundant. Therefore it can be concluded that a primary purpose of discounting is to allow for interest lost or payable. Discounting deals with the preference for early benefit receipt or late cost payment through the interest principle.

Investment return for use in the discount rate must be a mix of interest rates for both lost equity and borrowed funds utilised in the project. One issue that needs to be resolved, therefore, is the ratio of equity to borrowings, and this ratio must be applicable to all the cash flows that are expected to occur over the selected study period.

Often, borrowed funds are employed to finance up to 100% of the construction of the project. Operating costs are generally financed from equity reserves, which is a combination of commencing funds and progressively accumulated revenue. The proportion of borrowed funds in this case can be estimated as some or all of the capital cost divided by the total life-cost. The latter must be expressed in present value terms (i.e. not discounted value) and must exclude any consideration of finance costs. The proportion of equity is the remaining percentage so that together 100% of the life-cost of the project is considered. This type of information can be readily obtained from a budget or cost plan, or where not yet available it can be determined from an average of the proportion found in suitable past projects.

The proportion is not as critical as might be first thought. Interest rates on savings and interest payable on loans are not greatly different and generally follow similar trends. Therefore errors of judgement in arriving at a suitable proportion are not likely to be significant to the decision-making process. In any case, risk analysis needs to be undertaken on the discount rate and this process will clearly highlight the sensitivity of the discount rate in the context of the expected outcome.

RECOMMENDATIONS

The following conclusions and recommendations about the factors that influence the choice of discount rate used in a discounted cash flow comparison are listed below:

- The discount rate should be based on forecasts of investment return, inflation and taxation appropriate to the selected study period.
- Interest payable on borrowings or interest lost on use of equity should not be explicitly included in the cash flow forecasts but dealt with by the discount rate.
- Discount rates should be real (i.e. inflation should be removed) and the cash flow forecasts should be expressed in present-day terms.
- Taxation deductions, depreciation and liabilities should be explicitly included in the cash flow forecasts and the discount rate should be net after tax.
- Profit should not be built into the discount rate but its adequacy should be judged by the difference between the discount rate and the internal rate of return.
- Risk should not be built into the discount rate but should be assessed separately using one or more specialist risk analysis techniques.
- The discount rate should be calculated as the after-tax weighted average of equity and borrowed capital applicable to the individual projects being appraised.

'The higher the discount rate employed, the greater the weighting given to the earlier cash flows, so that projects that involve expenditure at an earlier date, but which produce savings in the future, such as preventative maintenance and energy insulation, appear unfavourably at higher discount rates.' (Grover and Grover 1987, p. 20)

The discount rate has been used in practice to account for a range of factors. It is argued that the loading of the discount rate is not an appropriate strategy. The outcome of this action is to distort the emphasis of initial costs and benefits relative to those that occur later, often resulting in the rejection of alternatives that avoid repetitive maintenance, reduce waste, save non-renewable energy resources, protect the environment or otherwise exhibit efficient operating performance.

Furthermore it is commonly stated in the literature that the discounting process discriminates against unborn future generations. Yet where the discount rate is based on the real weighted cost of capital, no discrimination is implied since the rate is merely allowing for the cost of borrowing money or losing interest on equity that otherwise would need to be considered as an explicit cash flow.

Discounting is a time-honoured institution. While modifications may be permissible on an incremental basis, wholesale rejection of the process is both unlikely and unwarranted. It is clear that some form of adjustment for time is necessary, but what is questioned is the manner in which this adjustment is determined and applied.

The discount rate is project-related, not market-related as is generally understood in the literature, and therefore may differ between competing investment opportunities. Furthermore, the use of artificially high rates of discount unreasonably distorts the equitable balance of initial and subsequent cash flows.

REVIEW QUESTIONS

3.1 What is meant by the term 'economic life'?

3.2 Why are accounting rate of return and simple payback methods of investment appraisal inappropriate?

3.3 Under what situations will multiple internal rates of return be calculated?

3.4 Why is taxation often ignored in discounting calculations?

3.5 What are the two types of taxation concession and to what type of expenditure do they apply?

3.6 How can the minimum acceptable rate of return be determined?

3.7 Why are finance costs ignored in discounting calculations?

3.8 Why are discount rates that include profit and risk undesirable as the basis of selection for alternative design decisions?

3.9 Why does discounting favour expenditure in the far future and revenue in the near future and why is this attribute of importance?

3.10 What are the two philosophies for discounting and why are they different?

TUTORIAL EXERCISE

Calculate the NPV, IRR and BCR for the following three investment options. Use a discount rate of 3% and a time horizon of 10 years. Which investment would you recommend (if any) and why?

Option 1:	
Year 0 capital cost	$100 000
Year 1-10 operating cost	10 000
Year 2-10 income	20 000
Year 10 sale	100 000

Option 2:	
Year 0 capital cost	500 000
Year 1-10 operating cost	50 000
Year 2-10 income	100 000
Year 10 sale	500 000

Option 3:	
Year 0 capital cost	250 000
Year 1-10 operating cost	10 000
Year 2-10 income	50 000
Year 10 sale	300 000

The above data are expressed in today's dollars and the discount rate of 3% is real (after inflation). The fact that the market value from the sale in Year 10 is equal to the capital cost is purely coincidental – market value could be more or less that what was originally invested and in many cases is a function of its capitalisation rate.

> **HINT**
> You may need to use a spreadsheet to test for IRR.

KEY POINTS

A PowerPoint presentation dealing with the topics discussed in this chapter can be downloaded from the publisher's website (see the publisher's details at the beginning of the book for the address). Some key points are shown below.

Cash Inflows and Outflows

* Discounted cash flow (DCF) analysis is a technique for assessing the return on capital employed in an investment project over its economic life

* Because cash flows that occur in different time periods have different values, they must be expressed in equivalent units

* Net benefits are discounted by an appropriate rate to calculate discounted present value

Investment Appraisal

* DCF analysis forms the basis for all investment appraisal techniques that take account of the changing value of money over time

* Two common applications of DCF analysis are net present value (NPV) and internal rate of return (IRR)

* NPis the sum of the discounted present values of all future cash inflows and outflows, and IRR is the discount rate that gives an NPof zero

Selection Criteria

* Projects are normally acceptable if their NPV is positive at a given discount rate

* The higher the NPV, the better

* IRR is the break-even discount rate and can also be used as a selection criterion

* If NPV is calculated with a discount rate based on the true time value of money, and if it is positive, then the difference between the IRR and the discount rate is a reflection of profit and risk

Taxation

* Taxation should be part of investment decisions

* Taxable expenditure can be divided into capital expenditure (which may be depreciable) and operating expenditure (which may be deductible)

* Depreciation is a fixed cost and therefore its value diminishes with the rate of inflation

* Tax deductions and depreciation should form part of the cash flow and the discount rate should be net after tax

The Choice of Discount Rate

* The discount rate may be based on the minimum acceptable rate of return for the project, the opportunity cost sacrificed by not investing in an alternative project, or the weighted cost of capital applicable to the financing of the project

* The latter is recommended

* Investment return is a weighted mix of the interest rate payable on borrowed funds and the interest rate lost on equity

Financing Implications

* The discounting process negates the need to include interest payable on borrowed funds and this is taken into account by the discount rate

* Discounting favours expenditure in the far future over expenditure in the near future by enabling the saved equity to theoretically earn interest

* The discounting process is not affected by the ratio of equity to borrowed funds, apart from the calculation of the discount rate itself

ANSWERS TO REVIEW QUESTIONS

3.1   Economic life is defined as the period of time during which the investment is the least-cost alternative for satisfying a particular objective.

3.2   Both techniques do not take account of the time value of money and therefore are simplistically assuming that money today and money tomorrow are equivalent.

3.3   Multiple IRRs may be calculated for projects that have a net cash flow with more than one change in sign (i.e. +/-) such as may occur in a multi-stage development involving construction, operation, further construction and operation.

3.4   Taxation is often ignored for reasons of simplicity, because the effects of tax concessions are not considered likely to change the outcome, or in public sector projects where tax is not applicable.

3.5   Depreciation allowances and tax deductions are both concessions against business-related expenses and relate to capital expenditure (acquisition costs) and operating expenditure (running costs) respectively.

3.6   Determination of the minimum acceptable rate of return may be based on a number of factors, but will typically reflect the interest cost of borrowed funds (preferably after tax) plus an allowance for profit and risk.

3.7   Finance costs, such as interest payable on borrowed funds and interest lost by use of equity, are part of the discounting process as this is the basis of the discount rate, and it is therefore not necessary to separately include interest costs in the cash flow.

3.8   While the inclusion of profit and risk may be acceptable in overall investment selection, it represents an artificial distortion of the true time value of money and will lead to incorrect decisions being made in life-cost studies.

3.9   Positive discount rates will reduce the importance of all future cash flows in relation to present cash flows and therefore favour early receipt of benefits and late payment of costs in the same way that society will favour them.

3.10 Time preference and capital productivity are rival discounting philosophies that produce a rate reflecting control, but the former is based on issues like impatience while the latter is based on interest calculations.

REFERENCES

Ashworth, A (1988), *Cost Studies of Buildings*, Longman Scientific & Technical.

Grover, RJ & Grover, CS (1987), 'Consistency problems in life cycle cost appraisals' in proceedings of Fourth International CIB Symposium on Building Economics, Copenhagen, pp. 17–30.

Peirson, G & Bird, R (1981), *Business Finance* (Third Edition), McGraw-Hill.

Robinson, JRW (1986), 'Life cycle costing in buildings: A practical approach', in *Australian Institute of Building Papers* (Volume 1), Paragon, Canberra, pp. 13–28.

Wright, MG (1972), *Discounted Cash Flow*, McGraw-Hill.

CHAPTER 4

LIFE-COST PLANNING

LEARNING OBJECTIVES
In this chapter you will learn about cost planning using a whole-of-life approach, including the types of life-costs that are relevant and suitable risk strategies that can be applied. By the end of this chapter you should be able to:
* understand the importance of life-cost modelling,
* describe the various forms of life-costs for a building project,
* select an appropriate study period, and
* apply risk management techniques to ensure that client recommendations are reliable.

THE PURPOSE OF COST PLANNING

Cost planning is part of the cost control process that commences with the decision to build and concludes with the completion of design documentation. During this period the main objectives are:

* the setting of cost targets, in the form of a budget estimate or feasibility study, as a framework for further investigation and as a basis for comparison,
* the identification and analysis of cost-effective options,
* the achievement of a balanced and logical distribution of available funds between the various parts of the project,
* the control of costs to ensure that funding limits are not exceeded and target objectives are ultimately satisfied,
* the frequent communication of cost expectations in a standard and comparable format.

The scope of cost planning should not be confined just to the construction of buildings, as is common in practice, but should include matters that are expected to arise during the life of the project. The process is aimed at improving value for money to the investor through comparison of alternatives that meet stated objectives and qualities at reduced expense.

The cost plan is one of the principal documents prepared during the initial stages of the cost control process. Costs, quantities and specification details are itemised by element and sub-element and collectively summarised. Measures of efficiency are calculated and used to assess the success of the developing design. The elemental approach aids the interpretation of performance by comparison of individual building attributes with similar attributes in different buildings, and forms a useful classification system.

Cost planning depends heavily on a technique known as cost modelling. Modelling is the symbolic representation of a system, expressing the content of that system in terms of the factors that influence its cost. Its objective is thus to 'simulate' a situation in order that the problems posed will generate results that may be analysed and used to make informed decisions.

Life-cost planning is similar in concept to capital cost planning except for the types of costs that are taken into account and the need to express all costs in common dollars. The aim is to prepare a document that describes the composition of the building in a manner that is of use to the investor or owner. A cost plan in this format could be used to demonstrate the relation between initial, replacement and running costs and to assist in the choice of specification and design details.

The objective of life-cost planning is not necessarily to reduce running costs, or even total costs, but rather to enable investors and building users to know how to obtain value for money in their own terms by knowing what these costs are likely to be and whether the performance obtained warrants particular levels of expenditure. A project is cost effective if its life-costs are lower than those of alternative courses of action for achieving the same objective.

Expenditure commonly associated with commercial type buildings includes acquisition, cleaning, energy, rates and charges, maintenance and replacement. These costs can apply to the building structure, its finishes, fitments, services and external works. Plant and equipment, in particular, requires technical advice on performance statistics and energy demands. The establishment of such life-costs may be difficult without historical data or expert knowledge.

Sidebar notes:

An *element* is defined as a portion of a project which fulfils a particular physical purpose irrespective of construction or specification. A *sub-element* is part of an element which is physically and dimensionally independent and separate in monetary terms. (NPWC 1980)

'If life cycle cost is to be effective, it must be implemented as early as possible in the design process ... the later life cycle cost techniques are introduced during the design process the lower will be the potential for cost savings and the more expensive it will be to implement any design changes suggested by the results of the analysis.' (Flanagan & Norman 1983, p. 11)

'The primary objective of a comprehensive methodology of life-cycle analysis is to maintain an up-to-date picture of options available to the owner and/or the user throughout the life cycle of a building.' (Bon 1989, p. 114)

The life-cost plan should be prepared on at least an elemental basis showing all quantities and unit rates. It should be set out in a fashion that enables extraction of totals of each type of cost category, including capital cost, along with costs per square metre and percentages of total cost. Life-cost planning, like any other form of cost investigation, is most effective in the early stages of design.

TYPES OF LIFE-COSTS

Life-cost planning applies to the acquisition and ownership of assets. The purpose of life-cost planning, as opposed to discounted cash flow analysis, is to determine the total costs of a building over a specified time frame in order to objectively assess the performance of the design in terms of durability, quality, energy usage and the like. Life-cost planning establishes estimated target costs for the capital and running costs of a building or building element. These targets provide a constraint and a useful measure against which selected design solutions can be later compared and evaluated.

Life-costs can be divided into various categories that aggregate capital and operating expenses in different ways. The following categories are used in this book as a basis for consistency:

- *Capital costs* Capital costs comprise the initial acquisition of the land and building, and can include:
  - *Land cost* The purchase cost of the land.
  - *Construction cost* The cost of labour, material and plant involved in the creation of the building and other improvements to the land, including all supervision, profit and rise and fall during the construction period.
  - *Purchase cost* Acquisition costs not directly associated with the finished product, including items such as stamp duty, legal costs, building fees, professional fees, commissioning and the like.
- *Operating costs* Operating costs comprise the subsequent expenditure required to service the land and building, and can include:
  - *Ownership cost* Regular running costs such as cleaning, rates, electricity and gas charges, insurance, maintenance staffing, security and the like.
  - *Maintenance cost* Annual and intermittent costs associated with the repair of the building, including periodic replacement or planned renovation.

'Denial-of-use costs ... include the extra costs or income lost during part of the life cycle because occupancy or production is delayed as a result of a design decision.' (Dell'Isola & Kirk 1983, p. 8)

- *Occupancy cost* Costs of staffing, manufacturing, management, supplies and the like that relate to the building's function, including denial-of-use costs.
- *Selling cost* Expenses associated with ultimate sale, including real estate agent commissions, stamp duty, transfer fees and the like.
- *Finance costs* Finance costs comprise expenditure relating to the interest component of loan repayments, establishment and account fees, holding charges and other liabilities associated directly with borrowed capital.

'The resources used to support buildings over their lives are as least as great, even in discounted present value terms, as the costs of their construction in the first place [and so] a total life view is therefore necessary for their performance to be properly planned and managed.' (Bromilow 1985, p. 151)

Historically there has been a serious bias against capital investment in favour of incurring running costs, which is obviously undesirable from the point of view of the general economic welfare. A reduction in initial construction costs often leads to higher maintenance and running costs.

Life-costs are of a significant nature and deserve consideration commensurate to their effect. Their absence from investment decision-making in the past has largely stemmed from the difficulties of applying these techniques and the tendency of building economists and building owners in particular to ignore the impact of long-term expenditure.

Full consideration of maintenance aspects and possible future costs at the design stage is likely to result in the building investor securing better value for money.

STUDY PERIOD

Study periods of sixty years are often quoted in the literature as reasonable. (Ashworth 1988, Seeley 1983, Stone 1980)

A life-cost plan is a 'blueprint' for the proposed expenditure of a construction project, or any asset, over its entire life span. Buildings commonly last a long time, in exceptional cases many centuries, before they collapse of their own accord or are torn down. The implication is that the life-cost plan must address itself to these costs in order to reach an optimal balance between initial and subsequent expenditure.

However, the factors that influence the termination of a building element or component are not solely those arising through physical deterioration, but can alternatively be influenced by inadequate design, workmanship or obsolescence. Premature damage through vandalism, adverse weather conditions and the like would have its cost implications offset by insurance cover.

The life of a building or other asset in the past has been particularly difficult to forecast because of premature obsolescence. This may be caused by one or more of the following:

The choice of study period methodology remains controversial.

- *Physical obsolescence* The physical life of the building is the period from construction to the time when physical collapse is possible. In reality, most buildings never reach this point as they are demolished or refurbished for other reasons.
- *Economic obsolescence* The economic life of the building is the period from construction to economic obsolescence, that is, the period of time over which occupation of a particular building is considered to be the least-cost alternative for meeting a particular objective.
- *Functional obsolescence* The functional life of the building is the period from construction to the time when the building ceases to function for the same purpose as that for which it was built. Many clients of the building industry, particularly in manufacturing industries, require a building for a process that often has a short life span. Functional obsolescence can also include the need for locational change.
- *Technological obsolescence* This occurs when the building or component is no longer technologically superior to alternatives and replacement is undertaken because of expected lower operating costs or greater efficiency.
- *Social obsolescence* Fashion changes in society can lead to the need for building renovation or replacement.
- *Legal obsolescence* Revised fire or occupational health and safety regulations and updated building ordinances may lead to legal obsolescence.

The study period can be set at between 25 and 40 years for most applications, particularly as eighty percent of ownership costs are believed to be incurred within twenty-five years. (Kirk 1979)

The life of an asset is commonly taken to be equal to its economic life. This is the period of time during which the asset makes a positive contribution to the financial position of its owners, both present and future. The only practical procedure is to work on the assumption that buildings are likely to be replaced at the end of this period. The concepts of both asset life and economic life, however, have disadvantages.

Once an owner of a building sells the property to another person, the original owner is no longer concerned with its operating performance or the costs or revenues that it will subsequently incur or generate. Hence the life of the building, or any other asset, should logically be only the period over which the owner has a financial interest. This duration then forms the time horizon or study period for all life-cost planning activities.

The concept of using 'financial interest' as the basis for assessing the time horizon of the study is not unique. As life-costing is concerned with the costs of ownership, the owner should ascertain and weigh costs from the time he/she considers acquiring the asset to the point when he/she intends to, or may wish to, dispose of it.

'It is often a good idea to set the study period equal to the period the building owner intends to hold the property, "the holding period". If this is done the assumption can be made in performing the economic evaluation that components lasting less than the study period are replaced and any residual values past the study period are captured as resale or scrap values (less disposal costs). Some project components may last less than the holding period and some longer.' (Ruegg & Petersen 1987, p. 21)

Others maintain that the time horizon is the period over which the building earns income or provides a service, for it is during this period, rather than the period for which it is required by the first owner, that the costs must be recovered. The general reason for this contradiction is a function of the perceived benefactor of the life-cost technique. Advocates of the 'financial interest' approach consider the investor as the focus of value for money pursuits, whereas those that support 'economic life' or 'asset life' consider maximisation of society's wealth as the objective.

Where a developer is constructing a building for immediate resale, it may be desirable to consider only purchase and capital costs. In this case the building has a zero life or a life equal to that of its construction period. The time horizon may vary depending on the type of investor. An investor may wish to consider the study period as being the holding period that is expected to maximise speculative profits. Once the building is sold, the financial interest of the developer is terminated and subsequent performance of the building is relevant only to the new owner.

Value for money to the investor, whether owner or developer, is the appropriate objective of life-cost planning. Interest in the total cost approach grew in Great Britain from the realisation that maintenance of their deteriorating building stock would become a significant social burden in the near future. But perhaps the lack of widespread adoption of this approach, particularly in the private sector, is in part due to past concentration on the theoretical benefits to society, which are often not consistent with those of the investor commissioning the study.

The selection of the study period remains a controversial matter in practice.

The literature on both discounting and life-cost studies advocates a study period based on either economic (useful) life or holding period. There appears no universal acceptance of one or other method. Further approaches to the determination of the study period, such as physical life and the analysis of premature obsolescence, are generally not given much regard.

Since the literature discusses life-costing primarily in the context of discounting, it is not surprising that the argument is frequently put that the study period becomes irrelevant after 25 to 40 years. The effect of discounting is such as to often render long-term costs and benefits negligible, so that there is little difference in outcome between a study undertaken over 25 years and one undertaken over 100 years or more. Discount rates employed in excess of around 5% per annum will generally have this effect.

At high discount rates the duration used for the study period has little consequence. However, as discount rates become lower or even negative, long-term costs and benefits take on more importance and the study period must capture these if a correct decision is to be made. Table 4.1 illustrates this effect for a $1000 payment or receipt.

Table 4.1 Value of $1000 at Various Discount Rates

Years	Discount Rate p.a.				
	-5%	0%	5%	10%	15%
0	1000	1000	1000	1000	1000
5	1292	1000	784	621	497
10	1670	1000	614	386	247
15	2158	1000	481	290	123
20	2790	1000	377	149	61
25	3605	1000	295	92	30
30	4659	1000	231	57	15
35	6021	1000	181	36	8
40	7781	1000	142	22	4

The issue of discounting and its effect on the study period is relevant only to comparisons between alternatives. Where measurement of a specific option is to be undertaken, whether it be for budgeting, cost planning or other purpose, the period of time over which the study is conducted is not affected by this phenomenon of 'diminishing returns'. A long-term investor, such as a government authority, may require the study to span an extended duration. But a maximum limit is required or else the study can never be completed.

It is further considered that the use of life-costing in the past has been impeded by the selection of the study period. For example, there is little attraction for a private sector investor to commission and pay for a life-cost study to be undertaken over 40 years where the investor is anticipating selling off the building within 10 years. Although there might be some marketing advantage on sale if the building can be objectively demonstrated to be energy efficient and of low-maintenance design, the investor is being asked to spend money now to benefit a future owner.

This highlights the fundamental dilemma between the often opposing objectives of investor interest and social welfare. While it may be conceptually correct to optimise a design so that society achieves the maximum benefit possible, this strategy may also result in no study ever being commissioned since the investor may be departing on a course that ultimately reduces his return. Issues of energy conservation and public sector investment suggest social welfare (economic life) as the most appropriate basis for the study period. Issues of profit maximisation and private sector investment suggest investor interest (holding period) as preferable. A solution is required that can support both perspectives.

The problem with the use of holding period is clearly that the interests of future generations may not be well served by the interests of the owner or developer, and hence decisions might be taken that are sub-optimal from society's perspective. The concept of sustainable development therefore appears inconsistent with a study period that focuses merely on the first owner. However, no study period (other than perpetuity) will ensure that the interests of future generations are protected. For example, a project that produces a high return over its economic life may well be a large user of non-renewable energy resources or may generate toxic waste that irreparably damages the environment even after the project ceases to exist.

The solution to this problem lies not with determination of the study period but with the use of a sustainability constraint. This entails debiting the project with the cost of resource depletion or environmental damage arising from its existence. This cost is converted into an annual equivalent based on the holding period and distributed accordingly in the cash flow forecasts. The mechanism to ensure that this process occurs must lie squarely with government regulation or legislation, as there is no incentive to disadvantage projects that would otherwise be profitable if environmental matters were ignored.

It is therefore considered on balance that the period of financial interest of the owner (holding period) is the appropriate study period for the investment. It will encourage the use of the life-costing technique and will be more relevant to the investor commissioning the study. Often this period is determined with reference to economic or useful life anyway, but it can differ markedly according to the objectives of the potential owner. For instance, developers may act as owners for a very short period and will usually wish to maximise their benefits rather than the benefits of one or more future owners. Maximisation of social welfare, however, does not need to be a mutually exclusive pursuit.

RESIDUAL VALUE

Knowledge relating to the expected timing of maintenance and replacement work can lead to the determination of the theoretical value of the building at the end of the period of financial interest of the owner. This value is known as the asset's residual value, and is essential in life-cost calculations to fully consider matters of durability and obsolescence. Residual value enables unequal asset lives to be reconciled, resulting in valid comparisons. This value can become quite important if one alternative requires major replacement towards the end of the selected time horizon.

Furthermore, it must be remembered that an asset can appreciate at a rate different from that of general inflation, due to changes in perception of its total value. Residual value should account for this enhanced (or diminished) growth in value through calculation of an asset appreciation rate, as follows:

$$a = \frac{1+g}{1+f} - 1 \qquad \text{(Eqn 14)}$$

where: a = asset appreciation rate p.a. (factor)
          g = expected growth p.a. (%) divided by 100
          f = inflation rate p.a. (%) divided by 100

A suitable approximation of the asset appreciation rate is $a = g - f$. This rate may, of course, be positive or negative. Replacement periods for individual components of the building can be used to calculate their residual life and hence residual value. For example, if a building was sold after 29 years and the life of the roof covering was estimated at 30 years, then the residual value of the roof covering can be described (using a straight-line approach) as one-thirtieth of its current capital cost. Performing this calculation for each component of the building would determine the residual value of the building itself. Care should be taken in the use of high appreciation rates coupled with long study periods, as the compounding effect can be significant.

The appreciation rate should not be confused with rates of specific inflation (i.e. escalation) applicable to various building components and systems. The component parts of the asset can collectively appreciate at a rate different from the specific inflation rates that would otherwise apply. Examples of situations where appreciation might be appropriate include assets such as computer installations, which tend to devalue rapidly as a result of technological advances and obsolescence, and assets that have antique value and hence may become worth considerably more over time due to their rarity.

'If the period of prediction for the building is limited to the period of expected occupancy by the first owner, allowance should be made for the residual value.' (Stone 1980, p. 59)

The importance of residual value is particularly highlighted where the period of financial interest applicable to the owner is used as the basis for time horizon in the life-cost plan. In cases where the building is not assumed to be demolished at the end of this period, the remaining potential performance (if any) will contribute significantly to the building's expected resale or market value, and is thus clearly relevant to the original investment decision.

PROBLEMS WITH TRADITIONAL TECHNIQUES

In recent years much publicity has been given to life-costing and its potential to quantitatively measure the performance of design decisions. While many advocate its concept, little evidence exists to suggest that this approach has been actively pursued by practitioners. One must stop and consider the reason why. Numerous problems (flaws) are continually highlighted in the literature. These possible flaws can be summarised as:

- *Accuracy* Life-costing relies on a number of assumptions, some of which are of a crucial nature requiring professional judge-ment, interpretation and experience of a high order. Matters such as the building's life, the lives of the various components used and their repair intervals and costs, rates for inflation and investment return and the effects of taxation all need to be predicted if the technique is to be used successfully in practice. A wide range of values can be allocated to all of these matters, which could result in the achievement of very diverse solutions.
- *Historical data* A frequently held reason for why the technique has not at present been more widely used is due to the lack of appropriate historical cost data. Where data have been found to be available they are so contradictory in their nature that their satisfactory reuse is almost impossible.
- *Maintenance policy* Predictions of cost based on theoretical durability and performance can be easily invalidated by the maintenance policy decisions of the future owner. Frequently maintenance work is undertaken in accordance with budget allocations on a priority basis, which may differ from that foreseen at the time of the estimate.

- *Professional accountability* The accuracy of life-cost estimates will not become apparent for many years, and even then may be difficult to compare to the costs of the day. Life-cost estimates are therefore hard to criticise and involve minimal account-ability on behalf of the building economist.
- *Predicting the future* Prediction of inflation and investment return levels for the following year is a difficult task. Predicting these levels with confidence well into the next century is impossible. Frequent changes to taxation policy further compound these unknowns.
- *Technological changes* The introduction of new materials and techniques possessing improvements over current technology can invalidate present decisions.
- *Fashion considerations* Often buildings are renovated for reasons of fashion and maintaining or improving rent levels than for repairing the effects of physical deterioration. Accounting for such renovation poses a serious problem, for although such changes may be gradual, their direction can be less predictable. The impact of work and leisure, personal expectations, demography and social development will all affect the way buildings will be used in the future.
- *Ownership of the building* Investors who do not intend to retain the building after its construction will find little value in life-cost calculations.
- *Financial resources* The financial position of the investor may dictate the need for minimum capital costs to be pursued. Perhaps the most important factor is a limit on the amount which the investor is willing, able or permitted to spend on building. This may be a major reason for the current failure to pursue life-cost studies.
- *Environmental factors* There is no guarantee that theoretically durable materials will significantly reduce maintenance costs, due perhaps to the effects of accelerated weathering or poor workmanship.

Clearly the most significant problem with traditional techniques is the discounting process itself. It is not just the selection of the discount rate that is problematic – the very nature of the adjustment is ill-defined. Since life-cost studies involve the consideration of costs over a number of years, the technique requires a method of bringing all costs into equivalent terms. So discounting has been adopted for this purpose, and with it has come the inherent difficulties.

But the identified problems are not so as to inhibit the use of life-cost studies, merely to impose constraints. Yet in the literature, where these problems are frequently listed, little active research is undertaken to overcome them. Instead, people seem to use these problems as reasons why the technique is not widespread in practice, whereas the real reasons may in fact be more a function of traditional industry procedure.

Even if all of the drawbacks were overcome, integrating life-cost planning into normal practice within the building industry would still be difficult. Two major reasons for this are the probable increase in professional fees and design time that would be required to prepare the study and the indifference of the typical building contractor to the life-cost plan.

RISK ANALYSIS

Risk is always present when making a decision but does not necessarily create a problem, especially where its impact is low. Even if its impact is high, risk can still be accepted by investors, because usually the greater the risk the higher the profit they expect to make. Investors must be able to perceive the presence of risks and accurately predict their magnitude and likely impact on the investment. This is accomplished by a systematic and disciplined approach to risk management.

Risk exposure is defined as the probability of investing in a project that has a less favourable economic outcome than the expected or desired target. Risk attitude is the willingness of the investor to 'gamble' on an investment of uncertain income.

The identification of risk and uncertainty is a central issue for any technique that involves the forecast of future events. Uncertainty (or certainty) refers to a state of knowledge about the variable inputs to an economic analysis, while risk refers to risk exposure or risk attitude. Methods available for quantitatively analysing these matters can be divided into two groups based on whether or not probability theory is involved.

Deterministic approaches to risk analysis include conservative benefit and cost estimating, break-even analysis, sensitivity analysis, risk-adjusted discount rates and certainty equivalent techniques. Probabilistic approaches are more complex and include input estimation using expected values, mean-variance criterion and co-efficient of variation, decision analysis, simulation and mathematical/analytical techniques. To be effective, uncertainty should be treated explicitly.

Sensitivity analysis is the most commonly employed risk analysis technique. It involves making minor changes to key variables in order to observe the effect on the originally predicted outcome. Risk and uncertainty can be minimised by demonstrating that the project is not 'sensitive' to such variations. It is dangerous to present results for any investment appraisal exercise without conducting a sensitivity analysis for a range of discount rates.

Sensitivity analysis can be used to determine the impact of risk by changing key variables in the project. For example, different discount rates in a reasonable range can be tested to determine the impact on net present value. Break-even analysis can also be used to determine the point at which the decision will change. Hence sensitivity analysis and break-even analysis are closely related and can well be considered as a single technique.

Another simple risk analysis technique is called conservative benefit and cost estimating technique. Using this technique the costs and benefits in a discounted cash flow are adjusted to reflect their uncertainty. For example, if a 5% conservatism adjustment is assumed, all costs will be increased by 5% and all benefits reduced by 5%, and the net present value is recalculated. The discount rate is not adjusted.

An alternative method for analysing the impact of risk is to load the discount rate with a risk premium. For example, the normal discount rate can be increased by 5% and the net present values recalculated. The effect of compounding results in an outcome at odds with the conservative benefit and cost estimating technique, as both deal with risk and uncertainty in different ways.

The risk-adjusted discount rate technique makes the implicit assumption that risk behaves in a compounding manner. However, doubt must be cast as to the appropriateness of this assumption. The discounting process is based on the principle of compound interest. The main purpose of discounting is to disadvantage future events over the present by consideration of the opportunity cost of money, so that costs and benefits in different time periods can be compared on an equivalent basis. The compounding effect is derivative of the rate at which money can grow given that it is invested and the proceeds are continually reinvested.

While there is reason for the discount rate to be applied to a series of costs and benefits in a compounding manner, the inclusion of risk in this rate is less justifiable. It is true, however, that higher discount rates reduce the impact of future cash flows, which are those of course with the greatest uncertainty. Nevertheless, the rate at which high discount rates reduce the value of costs and benefits leads to the proposition that risk, when included in this rate, may not be well represented by the compounding effect.

A simple comparison between a 5% risk allowance in the conservative benefit and cost estimating technique (i.e. costs increased by 5% and benefits reduced by 5%) gives quite a different result from a 5% risk allowance in the risk-adjusted discount rate technique (i.e. the discount rate is increased by 5%). The reason for the difference is the effect of compounding. While it is not possible to say that one approach is correct and the other in error, the risk-adjusted discount rate technique is regarded as less appropriate.

It is recommended that any technique that involves loading the discount rate artificially to include risk be avoided. A further reason supporting this position is that high discount rates usually result in the selection of low-capital-cost high-operating-cost alternatives that place an unreasonable burden on future generations and are likely to result in the accelerated usage of non-renewable energy resources. This outcome clearly has nothing to do with risk and is hence merely an unfortunate effect.

REVIEW QUESTIONS

4.1 What is the significance of the term 'cost targets'?

4.2 What is the purpose of cost modelling?

4.3 What are three approaches for the determination of asset life and which is preferred?

4.4 What does the term 'time horizon' mean?

4.5 What is the main advantage and disadvantage of orienting the time horizon of the life-cost plan to the period of financial interest of the owner?

4.6 What is the purpose of the asset appreciation rate and what is the pitfall in its use?

4.7 Are the difficulties associated with forecasting economic performance a deterrent to the implementation of life-cost planning?

4.8 What is the difference between deterministic and probabilistic approaches to risk management?

4.9 What other applications could residual value have other than comparative life-cost studies?

4.10 What can be done to maximise the benefits of life-cost planning?

TUTORIAL EXERCISE

Calculate the total life-cost of a school playground over 30 years given the following data:

Work Item		$ (2003)
A	initial capital cost	25 000
B	lawn moving and watering per chapter	25
C	lighting costs per quarter	150
D	playground repair per annum	500
E	repainting line marking every 5 years	1 000
F	resurfacing pavement every 10 years	10 000
G	replace seating/fitments every 10 years	5 000

In addition, graphically illustrate the percentage distribution of total capital to operating costs for the project.

The base year for the life-cost plan is 2003.

> **HINT**
> When measuring life-costs for a specific project (not compared to a range of other options), discounting is not relevant.

KEY POINTS

A PowerPoint presentation dealing with the topics discussed in this chapter can be downloaded from the publisher's website (see the publisher's details at the beginning of the book for the address). Some key points are shown below.

The Purpose of Cost Planning

❊ **The setting of cost targets as a framework for further investigation and a basis for comparison**

❊ **Identification and analysis of cost effectiveness**

❊ **Achievement of a balanced and logical distribution of available funds in the project**

❊ **Control to ensure funding limits are not exceeded and target objectives are ultimately satisfied**

❊ **Frequent communication of cost expectations in a standard and comparable format**

Classification of Life-Costs

❊ Capital costs comprise the initial acquisition of the land and building, and can include land, construction and purchase costs

❊ Operating costs comprise the subsequent expenditure required to service the land and building, and can include ownership, maintenance, occupancy and selling costs

❊ Finance costs comprise expenditure associated directly with borrowed capital

Asset Life

❊ Physical life is one approach to asset life that refers to structural integrity and safety

❊ Economic life is the period during which the investment represents the least-cost alternative for satisfying a particular objective

❊ The period of financial interest (holding period) relates to the intentions of the first owner

❊ Obsolescence governs asset life and may reflect any of the above considerations

Past Failings of the Technique

❊ Predicting the unpredictable

❊ Lack of a historical database

❊ Accuracy and accountability

❊ Maintenance policy of the owner/manager

❊ Technological and fashion changes

❊ Capital and operating budgets

❊ Environmental factors

❊ Conceptual problems with discounting

Risk and Uncertainty

❋ The identification of risk and uncertainty is a central issue for any technique that involves the forecast of future events

❋ Uncertainty (or certainty) refers to a state of knowledge about the variable inputs to an economic analysis, while risk refers to risk exposure or risk attitude

❋ Risk analysis techniques can be classified as deterministic or probabilistic

---

Documentation

❋ Life-cost planning begins with the production of a budget as a control mechanism for design

❋ Value management studies assist in the production of a sketch design life-cost plan which ensure the design is within budget

❋ Cost checks occur on a regular basis

❋ A tender document life-cost plan completes the design process and is the main cost management tool during the project's life

---

ANSWERS TO REVIEW QUESTIONS

4.1   The word 'target' is preferable to 'estimate' because the focus is not on predicting the future but setting up a framework within which effective cost management can occur, and should it be successful then the actual cost will be close to the original target.

4.2   Cost modelling is the simulation of the cost of a project before its construction commences, and refers to a wide range of activities and documents that occur throughout the life of the project.

4.3   Three common approaches to determination of asset life are physical life, economic life and period of financial interest (holding period) and all can be influenced by various types of obsolescence, but the period of financial interest is preferred in life-cost studies.

4.4   Time horizon is the study period or the period of time during which operating costs are considered, and it does not necessarily equal overall asset life.

4.5   The main advantage is that the project's design is being optimised for the person paying for the study to be undertaken; the main disadvantage is that social welfare may be inconsistent with investor wealth maximisation.

4.6   This rate is used to adjust the residual value of an asset, which may appreciate (or depreciate) relative to general inflation, but it should be used carefully as even small appreciation rates can compound to cause large differences in value over time.

4.7   Not really, because the proper use of risk analysis can enable the uncertainty of future events to be managed so that risk can be identified and form part of the decision-making process.

4.8   Deterministic approaches are based on prescriptive assumptions usually on a trial and error basis, while probabilistic approaches use probability theory to select ranges for variables and calculate the likelihood of their occurrence.

4.9   Residual value has uses also in insurance valuations for existing assets and in the preparation of depreciation schedules for taxation purposes.

4.10   Education of practitioners and marketing the benefits of life-cost planning to potential clients are the most important actions that should take place, but the technique needs to be supported by research into the benefits that can be expected over capital-centred evaluations.

REFERENCES

Ashworth, A (1988), *Cost Studies of Buildings*, Longman Scientific & Technical.

Bon, R (1989), *Building as an Economic Process*, Prentice-Hall.

Bromilow, FJ (1985), 'A life cycle approach to improving building performance' in proceedings of IDEA 85 Conference, Wellington, pp. 147–152.

Dell'Isola, AJ & Kirk, SJ (1983), *Life Cycle Cost Data*, McGraw-Hill.

Flanagan, R & Norman, G (1983), *Life Cycle Costing for Construction*, Surveyors Publications.

Kirk, SJ (1979), 'Life cycle costing: Problem solver for engineers', *Specifying Engineer*, vol. 41, no. 6, pp. 123–129.

NPWC (1980), *Cost Control Manual*, National Public Works Conference, Canberra.

Ruegg, RT & Petersen, SR (1987), *Comprehensive Guide for Least-Cost Energy Decisions*, US Department of Commerce, Washington.

Seeley, IH (1983), *Building Economics: Appraisal and Control of Building Design Cost and Efficiency* (Third Edition), Macmillan Press.

Stone, PA (1980), *Building Design Evaluation: Costs-in-Use*, E&FN Spon.

CHAPTER 5

LIFE-COST ANALYSIS

LEARNING OBJECTIVES
In this chapter you will learn about cost analysis using a whole-of-life approach and the importance of monitoring and managing existing buildings. Energy auditing is an interesting procedure in this regard. By the end of this chapter you should be able to:
- explain the difference between life-cost planning and life-cost analysis,
- describe procedures to monitor and manage buildings during their occupation,
- discuss the procedure of energy auditing, and
- use historical data pertaining to building performance in a meaningful way.

THE PURPOSE OF COST ANALYSIS

The collection and interpretation of actual building costs as an input or feedback source for cost planning is the main function of traditional cost analysis. It includes distributing detailed capital costs obtained from priced bills of quantities among the appropriate standard elements and sub-elements on which cost plans are normally based. Cost analysis has made possible a degree of control that was previously unknown. Composite unit rates can be averaged from historical projects to provide a valuable guide to the estimation of new projects.

The analysis of capital costs can logically be extended to encompass life-costs. Information concerning the running costs and performance of occupied buildings provides a vital source of data to both the building economist and the building owner. These data can then be used to highlight areas in which cost savings might be achieved in the design of new buildings, in the operation of existing buildings and in the choice of individual building components or systems.

Apart from providing feedback for future cost planning activities, analysis of life-costs can form an essential element of overall cost management by highlighting the ways in which potential cost savings in existing buildings might be achieved. For example, it may be better value to prematurely replace an expensive building component with a more efficient solution prior to the end of its useful life than to continue with a poor initial decision. Prudent control requires that actual and expected performance be constantly compared.

MONITORING AND MANAGEMENT

The monitoring and management of actual performance is herein referred to as life-cost analysis. While life-cost planning encompasses design cost control, life-cost analysis focuses on activities applicable to the construction and occupation of buildings. Together these techniques form the backbone of the cost control process and extend from the decision to build until the cost implications of the project are no longer of concern.

When considering initiatives for improvement, the option of 'doing nothing' should be regarded as a yardstick for comparison of other ideas. (RICS 1984)

In order for a life-cost approach to be effective in reducing the running costs of existing buildings it is necessary that these running costs be continually monitored. Monitoring involves the recording of actual performance of a particular project in a form that facilitates subsequent life-cost planning and management activities. Such performance can be compared against the cost targets and frequency expectations given in the life-cost plan. Areas of cost overrun or poor durability can be explored with an aim to implement identified improvements.

Management comprises the decision process and resulting actions that occur frequently throughout the building's life. It relies on the provision of meaningful data covering both cost and functional performance. Improvement initiatives are, to some extent, dependent on the maintenance policy of the building owner and available financial resources, but are vital if the objective of value for money is to be rigorously pusued.

DATA COLLECTION

Data collection is a prominent activity of life-cost analysis. Data from different projects will reflect the specific nature of the building, its location and occupancy profile. When measured overall, no two buildings will have identical running costs, nor will the running costs for any specific building be the same from one year to another. Data concerning past histories are not always directly applicable to the future but are nevertheless a necessary component of decision-making. Costs incurred at different points in time throughout the life of the building need to be converted back to a fixed project base date using known cost indices.

'If adequate monies are allocated for maintenance and service, the machine will never wear out.' (Coad 1977, p. 30)

A major problem when applying past cost information to new buildings is that the maintenance policy employed for existing buildings is not usually maintenance work will extend the lives of building components and delay the need for normal replacement or major service. The difficulty of assessing maintenance, however, can be partly overcome by averaging a large range of case studies for projects of similar function and specifying the actual costs in a simplified element-based format. Simulation techniques may also be of advantage.

In an environment of changing technology such as the construction-related industries, historical cost data must be interpreted considering their contemporary technologies. This is of importance when the implications for the design of new facilities as well as the redesign of existing facilities is fully considered.

Accurate life-cost studies 'cannot be undertaken without reliable and comprehensive data support, both in the area of cost and the physical aspects of buildings and its components.' (Rosenbauer 1986, p. 4)

Frequently one of the main aims of data collection is to help identify the characteristics of physical assets that have the greatest impact on total costs. These characteristics are often termed the main 'cost drivers' and should receive the largest part of management consideration in order to optimise the life-cost.

Data collection of building performance is still in its infancy across the existing stock of buildings. This absence of data is closely associated with the low level of life-cost planning currently evident in practice. But even in the United States, where life-costing is more popular and in some cases compulsory, reliable data from existing buildings are sparse. There are at least four reasons for this:

- The variable durability of materials and systems and the effects of their interaction with each other, the environment and the actual users of the building lead to a wide range of results in the data.
- The long time frame involved in the measurement of component lives and maintenance intervals makes accurate record-keeping vital yet problematic.
- The inability of building owners to compile data in an appropriate and available form for building economists inhibits its effective reuse on subsequent projects.
- The quality of the original construction directly affects performance.

Data from existing buildings should be viewed as an additional source of knowledge on which life-cost planning can draw, but should not be considered as the primary source. Estimates of operating costs and life expectancies can be made in a similar manner to that undertaken with capital costs by obtaining basic information from suppliers, maintenance sub-contractors and testing authorities. Building economists can also use knowledge of life-cost performance acquired from their own experience to assist with predictions.

The future of data collection may lie with the proliferation of buildings that continually monitor their own performance and provide detailed statistical feedback to the building owner and ultimately the building economist. This may be achieved through sensors and other mechanisms mounted in the building fabric and linked electronically to a controlling computer. Apart from providing relevant empirical data, 'intelligent' buildings will offer greater insight into the deteriorating effects of age.

CLASSIFICATION SYSTEMS

Life-cost data can be collected from existing buildings or estimated directly from first principles. In either case a suitable classification system is required. Elements and sub-elements are the most appropriate classification system for life-cost analysis. In this way collected data can be readily compared to the information predicted at the cost planning stage.

This strategy lends itself to computerised data retrieval systems. Life-costs need to be date-indexed so that adjustment using cost indices can occur. In addition, all stored costs should exclude the head contractor's allowance for preliminaries, profit and general overheads so that this aspect can be separately assessed (i.e. all stored costs must be equivalent to conventional sub-contractors' prices).

Life-cost data must be stored in an appropriate form. Sub-elements can be used to record the following:

Maintenance work is undertaken in an isolated manner at different time intervals on an individual contract basis. Any supervision and overheads on behalf of the building owner are best considered as a separate ownership cost.

- *Construction cost* This is the unit rate for initial construction or acquisition of the item ($ per sub-element unit).
- *Annual cleaning costs* These costs relate to regular cleaning activities ($/year per sub-element unit).
- *Annual energy costs* Energy costs concern regular expenditure on electricity, gas, solar and fossil fuels, but exclude the energy impact created by the relationships with other building components and systems ($/year per sub-element unit).
- *Annual repair costs* These comprise the cost of regular servicing and repair work, excluding owner supervision ($/year per sub-element unit).
- *Intermittent repair period and costs* This represents the interval for major maintenance work and the associated costs (years and $/period per sub-element unit).
- *Component life* This is the time during which the item is expected to remain in service (years).
- *Replacement costs* This cost is essentially the item's original construction cost plus an allowance for demolition or removal of the item ($ per sub-element unit).

This level of detail is sufficient for cost planning purposes, including budget preparation, estimating and making informed decisions about design alternatives. Recording useful knowledge about the performance of the building is an obvious application, but the challenge for life-cost analysis is more to supplement conventional estimating processes with data extracted from a wide range of existing buildings of similar function and supported by computer-based statistical inference techniques.

ENERGY AUDITING

The sustainable development philosophy implicitly suggests that new development should be designed so as to minimise its impact on the consumption of resources required for construction and future operation. While some control may be able to be exercised over this matter within normal investment appraisal activities, it is argued that additional control is necessary. The process of energy auditing can provide this control, but to be successful it must become a mandatory part of development approval.

Energy auditing can be divided into two parts: estimation and verification. Estimation involves the establishment of energy targets based on the design of the project and projections of the level of occupational usage. These would initially concentrate on operating performance but may in the longer term be extended to include the energy implied in resource extraction, processing and manufacturing. Verification involves the monitoring of actual energy use and comparison against the original targets. Action may then be necessary to correct situations where energy usage is shown to be excessive.

Life-cost studies will have an important part to play in the future to ensure that sustainability is achieved. Government initiatives in the area of energy auditing, while still embryonic, are set to grow so that eventually all new buildings will have to formally demonstrate that they are energy efficient. The amount of energy implicit in the manufacturing and construction processes may also come under new scrutiny, and may require building products to be 'rated' in much the same way as electrical appliances or car fuel economy are currently treated.

Energy auditing can be integrated into the development approval process with relative ease, since planning authorities already have the jurisdiction over whether new development is acceptable. Each development application should be accompanied by a statement of annual energy expectations appropriately classified into type and cause.

But verification is a different kind of problem. It requires an independent audit of actual performance in much the same manner as an accountant might verify the accuracy of financial statements and taxation liabilities. Therefore an annual submission to a central authority setting out total energy consumption is necessary. The building owner has the opportunity to balance the total energy usage by making savings in some areas to compensate for overruns in others.

All development applications would need to substantiate that their energy targets are within the limits set down by the regulatory authority in order that approval be granted. This can be achieved using a life-cost approach based on element and sub-element classifications, but rather than exploring the targets in terms of present value they would be expressed in units of energy (such as kWh or a suitable alternative). Failure to meet established energy targets would attract a financial penalty for the building owner that could be collected as part of the taxation system. Existing buildings would also need to comply for reasons of equity and fairness, but perhaps would be given dispensation through higher energy limits.

Life-cost planning and life-cost analysis have obvious similarities to energy auditing activities and hence it is logical and desirable that both be integrated as much as possible. For example, life-cost plans would state energy consumption units against relevant items as part of the measurement of present value. Annual activity reports would identify the actual consumption for the year at the same time that cash expenses are recorded. Management action plans would be able to rectify situations of excessive consumption by establishing policy such as a reduction in operating hours, replacement of old or poorly performing equipment or the implementation of an energy-saving campaign.

REVIEW QUESTIONS

5.1   What is the purpose of life-cost analysis?

5.2   Why do we need cost information from existing buildings?

5.3   Can management of a facility occur without the monitoring of performance?

5.4   When investigating new solutions to existing designs, what alternative should always be considered?

5.5   How can the maintenance policy of the building owner cause life-cost predictions to be inappropriate?

5.6   Can we life-cost plan without historical data?

5.7   What are the reasons for the low level of life-cost data collection in the past?

5.8   What is meant by the term 'intelligent building'?

5.9   What advantages can intelligent buildings offer to the process of life-cost analysis?

5.10   What is the problem with a sub-elemental approach to data collection?

TUTORIAL EXERCISE

Refer to the tutorial exercise from Chapter 4. It is now 2013. Your task is to determine whether the playground project has performed within the original expectations set out in the life-cost plan.

Actual expenditure for the past 10 years has been recorded and is shown in the table on the next page.

Year	Plan (PV) ($)	Actual (FC) ($)	BPI
0	50 000	48 785	100.0
1	2 400	2 505	102.1
2	2 400	2 490	102.7
3	2 400	2 500	103.6
4	2 400	2 610	104.0
5	3 400	3 560	105.2
6	2 400	2 650	106.9
7	2 400	2 700	108.1
8	2 400	2 700	110.3
9	2 400	2 750	110.2
10	18 400	20 500	111.8

HINT
To adjust for inflationary movements, work out the per cent change between the BPI (building price index) for a particular year and the BPI for the base year (2003 in this case).

KEY POINTS

A PowerPoint presentation dealing with the topics discussed in this chapter can be downloaded from the publisher's website (see the publisher's details at the beginning of the book for the address). Key points are shown below.

The Purpose of Cost Analysis

❋ Cost analysis has traditionally distributed actual costs among the appropriate elements (or sub-elements) on which cost plans are normally based, and has made possible a degree of control that was previously unknown

❋ Composite rates can be averaged from real projects to provide a valuable guide to the estimation of new projects

❋ Life-cost analysis is essentially no different

Data Collection

✲ Data collected from different projects will reflect the specific nature of the building, its location and occupancy profile

✲ When measured overall, no two buildings will have identical running costs, nor will the running costs for any specific building be the same from one year to another

✲ Obtained results should be applied to other situations with caution

Problems with Data Reuse

✲ Variable durability of materials and systems and the effects of their interaction, the environment and users of the building will lead to a wide range of results in the data

✲ Long time frames result in an incomplete picture

✲ Inability of building owners to compile data in an appropriate form inhibits their reuse

✲ Quality of the original construction directly affects performance

Classification Systems

✲ An elemental or sub-elemental approach is generally preferred as the basis of any classification system for life-costs

✲ This requires consistent standards to be applied across projects so that data are comparable

✲ Until an adequate source of data is available, or even after it is available, estimation of life-costs from first principles will be more reliable than averaged historical performance

> # Documentation
>
> ❊ **Activity reports are prepared at the end of each year of operation and collect all the expenditures that have occurred over this time, quantity information and reasons for the work**
>
> ❊ **Management action plans are prepared when an opportunity for improvement is discovered and it is economic to make a change**
>
> ❊ **Value management studies can assist in this process**

ANSWERS TO REVIEW QUESTIONS

5.1   Life-cost analysis is aimed at collecting information concerning actual performance and its interpretation during the construction and occupancy stages of a project's life cycle.

5.2   The collection and interpretation of actual life-cost performance is used as a feedback source for future design processes and as a part of the ongoing management of the existing facility.

5.3   No, since a proper decision-making process requires knowledge about expected and actual performance in order to deliver further benefits over the project's life.

5.4   The 'do-nothing' option, or in other words the existing situation, should be used as a base for any comparison with new initiatives.

5.5   There are probably many reasons, but the most significant one is the constraint imposed by limited funds and the consequent deferral of routine work, resulting in accelerated aging or increases in future cash needs.

5.6   Yes, in fact we must, since the level of historical data is poor and estimates therefore need to be built up from first principles in most cases.

5.7   The variable durability of materials and systems and their interaction with each other and their environment, the long time frame involved, the inability to record data in an appropriate format and the effect on performance of the quality of the original construction.

5.8   Intelligent buildings have technology built into them to record performance, to control service systems and to enable effective communication.

5.9 Technology can be used to automatically collect and interpret actual performance through the use of sensors and meters and to control systems so that they are tailored to the needs of the building users at various times during the day.

5.10 Collected data are so context-dependent that averaged costs/m^2 per sub-element are of little value unless there is attached information regarding location, occupancy profile, occupancy hours, etc. and this can be used to select appropriate cost records.

REFERENCES

Coad, WJ (1977), 'Investment optimization: A methodology for life cycle cost analysis', *ASHRAE Journal*, vol. 19, no. 1, pp. 29–33.

RICS (1984), *Life Cycle Costing: A Worked Example*, Surveyors Publications.

Rosenbauer, R (1986), 'Life cycle costing: The macro view' in proceedings of Building Science Forum of Australia (NSW Division) Seminar, Sydney, pp. 2–6.

CHAPTER 6

CHANGE IN WORTH

LEARNING OBJECTIVES
In this chapter you will learn about the concept of change in worth and its application to the discounting process. Differential price level changes and diminishing marginal utility collective contribute to future values. By the end of this chapter you should be able to:
- describe the determinants of change in worth,
- discuss how changes in living standards can affect affordability of goods and services in the future,
- recognise that discounting should additionally take account of changes in worth, and
- use formulae to adjust present values to future values.

BASIC WORTH

Equivalence in the assessment of costs and benefits depends on present and future generations valuing money in the same way. The discounting process uses interest rates to make this translation. This is not an unreasonable action but may over-simplify the time value of money because two further considerations are being ignored:

- *Differential price level changes* Goods and services, whether they be costs or benefits, may escalate or de-escalate at a different rate from general inflation. Thus their perceived value in the future will change relative to their value today.
- *Diminishing marginal utility* Goods and services, whether they be expressed as costs or benefits, may become easier or harder to afford in the future depending on individual prosperity. This will result in changes in their rate of consumption and hence their perceived value.

These considerations combine to indicate relative changes in the base worth of traded goods and services over time. The discounting process must take account of escalation and affordability if a proper comparison between present and future events is to be made. Each of the above matters will now be investigated in further detail.

DIFFERENTIAL PRICE LEVEL CHANGES

The use of differential escalation rates 'is suspect for prolonged investment periods'. (McGeorge 1988, p. 4)

The rationale for working with value rather than cost is reinforced by the relativity that commodities display with the general inflation rate. This relativity, however, is potentially upset by differential price level changes (drift). While various rates of specific inflation, or escalation, are likely to occur in the short term, such as between energy-, labour- and material-based costs, over longer periods these differences are likely to diminish. Differential price level changes would need to be large to affect the order of life-cost comparisons.

Therefore, ignoring escalation may be the result of it not being a significant issue in practice. But this situation may change over time and is likely to have different characteristics in different countries depending on access to natural energy reserves. The future cost of limited non-renewable energy resources is a probable candidate for rates of escalation exceeding inflation over the longer term.

'If all costs can be assumed to escalate at the same rate no great difficulty will arise ... [but if] different cost elements typically escalate at different rates, a more complex set of techniques will be necessary.' (Flanagan & Norman 1983, p. 19)

Changes in escalation over time can be expressed in the form of an escalation rate. By definition, therefore, escalation rates may differ from the general inflation rate, potentially resulting in a large number of different exchange rates. In the context of life-cost studies this range can be conveniently condensed to cleaning, repair, replacement and energy categories. Forecasts need to be made for escalation rates in each of these four cost types, and can best be achieved by the analysis of historical trends. Ultimately it is the differential price level change between escalation and inflation that is important.

Calculation of differential price level changes can be made by increasing present value by the applicable escalation rate and reducing the resultant figure by the inflation rate. If expected escalation equals inflation, no differential price level change will occur.

Cleaning costs and repair costs are primarily labour. Replacement costs are more a combination of labour and material. Energy costs are a function of a number of commodities such as crude oil, refined petroleum, coal, uranium, natural gas and the like. It is considered that non-renewable energy resources are likely to increase in cost as they become scarce.

Costs and benefits involved in investment projects can additionally escalate in a different manner from general inflation due to market-related factors such as consumer confidence, unemployment levels, competition, economic growth and the like. The construction industry is classically cyclical, moving from buoyant to recessionary conditions in an inevitable yet unpredictable manner. In times of recession the profit levels of contractors are reduced and although labour and material costs may continue to increase the overall price of construction can fall. In 'boom' times contractors' profit levels are adjusted to make up for past losses, and construction prices tend to rise at an accelerated rate.

In a similar way items such as rental income can be affected by the economy. Vacancy rates and a tendency towards decentralisation away from expensive inner-city premises can result in building owners being forced to keep rents as low as possible to maintain existing tenants and attract new ones. This may have the effect of initiating renovation work, even though it may not be physically necessary, in order to increase the attractiveness of the rental accommodation compared to other offerings in the marketplace. Investment projects that deliver services to the community can also have their costs and benefits influenced by supply and demand fluctuations.

Therefore market conditions contribute to the economic appraisal and subsequent performance of investment projects, and need to be carefully considered. This can be achieved either explicitly or implicitly. Explicit consideration may comprise adjusting costs and benefits in future years according to expectations of economic performance. This is of course very difficult for anything beyond the short- to medium-term. Implicit consideration may comprise use of risk analysis techniques to model the impact of market condition trends.

Since market conditions are cyclical and their frequency largely unpredictable, it is recommended that risk analysis be the preferred approach. While it is certainly true that the base worth of goods and services will be affected, over the longer term these fluctuations are theoretically absorbed by the escalation rate and hence need not be further considered in the discounting process.

International exchange rates are a further aspect of escalation. Projects constructed and operated offshore will be subject to movements in exchange rates between the home and host country. Local projects that have a significant content of imported goods may be similarly affected. Exchange rates are not cyclical and thus are more appropriately considered by the escalation rate. Therefore international exchange rates can be incorporated in the formulation of the discount rate for relevant projects.

DIMINISHING MARGINAL UTILITY

Utility is the satisfaction that an individual receives from the consumption of goods and services and marginal utility is defined as the extra value obtained or obtainable from the acquisition of one more unit of consumption. The elasticity of marginal utility is the percentage change in marginal utility resulting from a 1% increase in consumption. As increases in income enable further consumption to occur, marginal utility can alternatively be described in income terms.

Diminishing marginal utility requires four postulates of consumer behaviour to hold: (1) each consumer desires a multitude of commodities and no one commodity is so precious that it will be consumed to the exclusion of all others, (2) consumers must pay prices for the things they want, (3) consumers cannot afford everything they want, and (4) consumers are rational and seek the most satisfaction they can get from spending their limited funds for consumption.

In a world supposedly typified by advancing technology and growing affluence among its citizens, the increasing abundance of products leads to reductions in their marginal utility. This process can occur in two ways. First, there is consumer ranking of applications for available resources and products. Where scarcity prevails these resources and products are first used for their most important purposes, but where they are abundant their application is more trivial. Second, if more of the resource or product is used for the same purpose, diminishing returns in production set in and satiation in production may be approached. Furthermore, as the chief benefit of income lies in its power to buy resources and products, the marginal utility of increasing income also diminishes.

The principle of diminishing marginal utility is that the more of a good or service that is consumed, the smaller is the marginal utility obtained by consuming one additional unit. Dividing the change in total utility by the change in quantity consumed calculates marginal utility, which will exhibit a diminishing trend as the quantity consumed rises. Goods and services can be substituted for each other so that maximum utility is achieved within the constraints of available income.

This is known as the substitution effect. It is the change in the quantity demanded of a good or service that results only from a change in the relative price of the good or service. It differs from the income effect, which is the change in the quantity demanded of a good or service that results from a change in real income.

Equilibrium is created that implies that the marginal utility per dollar obtained from the last unit of each good or service consumed is equal for all goods and services.

Diminishing marginal utility is arguably the single most important concept in economics. It underlies the downward slope of demand curves. It is the most important basis for quantitative decisions on welfare distribution between individuals and groups in society. And correctly understood, it yields a cause for discounting that offends fewest cultural, philosophical and religious norms, while allowing a flexible interpretation of reality.

Assertions that succeeding generations will be more affluent than their forebears are not necessarily beyond dispute, even in cases where real income may increase. The cost of the 'standard of living' may also be proportionally higher, a fact not generally considered in the literature. Nevertheless, the general presumption of increasing affluence being a justification for discounting remains.

Affordability is a measure of changes in the purchasing power of money over time and can be expressed in the form of an affordability rate. It can apply to both costs and benefits, indicating that costs may become harder or easier to afford in the future and benefits may be of more or less assistance relative to current values. A positive affordability rate indicates a rise in purchasing power, thus making costs easier to afford and benefits of more assistance.

The concept of affordability is employed to measure the purchasing power of real income and hence to determine its marginal utility. The use of purchasing power rather than income itself is made to account for distributional changes in prosperity. For example, while an individual's real income may rise by 1% per annum each year, general economic growth may rise at 2% per annum. Thus, although more goods and services are able to be consumed with the extra income, the individual is not better off as his/her standard of living has fallen relative to that experienced by others in the community. It is this concept that enables

Affordability is concerned only with cash flows of a tangible nature. Increases in prosperity imply improvements in living standards, which are translated into the ability to acquire more goods and services per unit of income. Expenditure and income analysis is therefore crucial to the measurement of affordability changes.

changes in intergenerational equity to be linked to conventional utility theory.

Housing affordability is commonly advanced as a major factor in the assessment of changes in the living standards of consumers. In situations where consumers are being priced out of the housing market, perhaps by high land values or by high interest rates on mortgages, living standards are identified as being in decline. Housing affordability therefore can and has been used to monitor the success of economic policy and ultimately the performance of governments.

LIVING STANDARDS

A better approach to the measurement of intergenerational equity change is therefore to explore the relationship between future prosperity and living standards. If future generations are expected to be better off than the present generation it can be concluded that living standards improve. Living standards and quality of life are generally regarded as synonymous. Intergenerational equity is about ensuring that living standards do not deteriorate over time as a result of current action.

Commonly economic conditions are targeted in isolation when discussing living standards, as they form a tangible and objective measure of general performance. In many cases income and expenditure comparisons are used as the basis for this discussion.

There remains no single definition of living standards upon which to base its measurement. (Zwartz & Marcus (1989)

The common difficulty found in measuring changes in living standards is due to the imponderances of social issues pertaining to the quality of life. Working hours and conditions, improvements in technology levels for manufactured goods and services, greater variety of choice in purchase, attainment of lifestyle objectives, law and order and environmental issues are just a few of the facets of living standards that will affect and defy meaningful assessment. Living standards are judged with reference to the individual. The performance of governments and corporations is reflected by the benefit they ultimately provide to their citizens and shareholders. Final analysis must be a function of the perception of individuals in comparison to the standards being achieved by others.

Living standards form a suitable methodology for the measurement of affordability. Four main approaches adopted by researchers in various parts of the world are identified and categorised as follows:

- *Income and expenditure approach* The measurement of living standards using income and expenditure patterns has formed a mainstream of research effort. Real disposable income, if increasing, is considered to lead to corresponding increases in living standards. Yet others conclude that looking at income alone does not account for the use of savings and debt and believe that actual consumption is a more meaningful indicator.
- *Relative deprivation approach* This approach focuses on poverty. If individuals are deprived of the essentials of life then compared to others in society their deprivation level increases and their living standard conversely falls.
- *Level of living approach* This approach concentrates on the individual's command over resources as opposed to needs satisfaction. Resource has a broader meaning than merely real disposable income. A high level of living is achieved when individuals have a wide choice of actions on which to draw to shape their lives and living conditions.
- *Quality of life approach* Increases in material affluence are seen as contributing to but not fully explaining improvements in well-being or quality of life. Quantification of the worth of surroundings and lifestyle objectives, or in other words the values that an individual might seek to achieve, at different stages in an individual's life cycle, are the basis for measurement of living standards under this approach.

Attempts to measure living standards in the past have been hampered by the introduction of such complexities as different classes of individuals in society, regional effects and the impact of various wealth profiles. Yet it is the assessment of subjective concepts that has proven to be the most elusive and controversial.

Quality of life is considered the global term for lifestyle satisfaction, including all the aspirations and desires for the future, and the standard of living is seen as a subset based solely on materialistic criteria. The standard of living can thus be defined to encompass only matters of a monetary nature and measured by the value of goods and services held by individuals. The more goods and services people possess, the higher is their standard of living. Thus in terms of quality of life the objective factors are effectively separated from the subjective. While individuals might exhibit different standards of living, it is the rate of change that occurs that is perhaps of the most interest.

By defining the standard of living as a subset of the quality of life, the value of goods and services acquired by individuals for their own use enables changes in the standard of living to be identified. Furthermore, if this trend is compared with the level of real income that individuals earn then a measure of the ability to afford these goods and services is derived.

As the value of acquired goods and services grows in real terms per person, so does the resultant standard of living. Yet increases in expenditure without corresponding increases in income might suggest that people are living beyond their means, perhaps through reliance on credit or by running down their accumulated savings. Thus an affordability gap can be created. Affordability is a function of an individual's 'access' to goods and services.

FUTURE VALUE

Present value is defined as the worth of goods and services today. In other words, present value is the worth of goods and services to the present generation. Present value ignores changes that occur in base worth over time.

Future value is defined as the worth of goods and services at some subsequent point in time and is determined by the consideration of escalation and affordability. Future value is therefore the worth of goods and services to future generations.

Assuming that no change in base worth is expected to occur, then the value of future costs will not change and future value will be equal to present value. Where escalation and affordability changes result in an expectation that goods and services will become easier to obtain, the present value of future costs needs to be reduced. Future value is therefore less than present value in real terms. Where escalation and affordability changes result in an expectation that goods and services will become harder to obtain, the present value of future costs needs to be increased. Future value is therefore greater than present value in real terms.

When considering the value of future costs incurred over a specified life, the present value of anticipated cash flows must therefore be adjusted for the expected change in worth. The following formula is used for this purpose:

$$w = \frac{(1 + f)(1 + r)}{1 + e} - 1 \qquad \text{(Eqn 15)}$$

where: w = change in worth p.a. (factor)
f = inflation rate p.a. (%) divided by 100
r = affordability rate p.a. (%) divided by 100
e = escalation rate p.a. (%) divided by 100

A value for *r* of 0.015 would indicate that future generations are expected to find purchase of goods and services more affordable than present generations at a rate of 1.5% per annum.

A suitable approximation for describing the change in worth is w = f + r − e. It must be kept in mind that the expected change in worth used is assumed to remain constant throughout the life of the investment, and prediction over more than a few years can only be an assessment based on current trends. A positive affordability rate indicates that the cost of goods and services is becoming easier to afford and/or the purchasing power of available income is increasing. Change in worth is used as a measure of time preference.

The value of planned expenditure or revenue to future generations arising during the study period can be conveniently measured and expressed in 'equivalent' dollars using derivatives of Equations 8 and 9:

$$FV = PV(1 + w)^{-n} \qquad \text{(Eqn 16)}$$

where: FV = future value ($)
PV = present value of single payment or receipt ($)
w = change in worth p.a. (factor)
n = number of years

and

$$FV = \frac{PV[1 - (1 + w)^{-n}]}{w} \qquad \text{(Eqn 17)}$$

where: FV = future value ($)
PV = present value of annual payment or receipt ($)
w = change in worth p.a. (factor)
n = number of years

REVIEW QUESTIONS

6.1   What is the essence of the affordability approach?

6.2   Can changes in living standards be reasonably measured?

6.3   What is the meaning of a negative affordability rate?

6.4   What adjustments have to be made to the actual expenditure and income data that are collected year by year in order to draw conclusions about trends?

6.5   A decline in the ability to afford goods and services may not necessarily affect the standard of living for what reason?

6.6   Is it reasonable to use national aggregate figures about performance and then apply them to an individual person or project?

6.8   What is future value? Is it real?Q7Is it reasonable to consider that public and corporate sectors can be evaluated using the standard of living model?

6.9   Why does the formula for change in worth ignore interest and taxation?

6.10   How can we deal with the value of subjective considerations?

TUTORIAL EXERCISE

Determine the total expected future value for car ownership over 10 years based on the following cost profile. Assume inflation is 5%, escalation is equal to inflation except for fuel costs (7%), and affordability change is –1% each year.

- purchase: $38000
- fuel costs (annual): $1500
- maintenance costs (annual): $3000
- new tyres (every 5 years): $600
- new radiator (every 7 years): $250
- new exhaust system (every 10 years): $500
- resale after 10 years: $5000

KEY POINTS

A PowerPoint presentation dealing with the topics discussed in this chapter can be downloaded from the publisher's website (see the publisher's details at the beginning of the book for the address). Some key points are shown below.

Affordability Approach

❋ **Traditional discounting takes explicit account of the cost of finance, and therefore leads to the determination of comparative value using an investment–based or opportunity approach**

❋ **The value of future costs and benefits is also susceptible to fluctuations in basic living standards, and the resultant change in worth leads to the determination of comparative value using an income–based or affordability approach**

Standard of Living

❊ Lifestyle satisfaction can be classified as comprising standard of living (objective) and quality of life (subjective) considerations

❊ Standard of living represents the ability of individuals to purchase goods and services

❊ As the value of acquired goods and services grows in real terms so does the resultant standard of living

❊ It is an expectation that standards increase

Affordability Indices

❊ While standard of living may increase, the ability of individuals to afford such a standard may fluctuate

❊ Comparison of expenditure and income patterns will reveal whether increases in standards are matched by increases in income earned, and the results can be expressed as an index

❊ Differences between the expenditure and income contribute to either wealth or debt accumulation

Affordability Rate

❊ Changes in the ability to afford increases in the standard of living over time are represented by the affordability rate

❊ A positive affordability rate means that things are becoming more difficult to afford, while a negative rate means that things are becoming easier to afford

❊ It is likely that the affordability rate will approach zero when measured over long time horizons

Differential Price Level Changes

�֍ Specific inflation (or escalation) will lead to differential price level changes over time

✷ The rate of change for a particular commodity is a function of the difference between general and specific inflation

✷ Escalation can be categorised generically as energy-, labour- or material-based, or can be treated as product-based

✷ Technology change can be allowed for in this way

Change in Worth

✷ Affordability and escalation changes combine to determine changes in the base value of goods and services

✷ Change in worth is important when making comparisons between alternatives

✷ Present value needs to be adjusted for change in worth

✷ Ignorance of changes in worth are an over-simplification and may lead to significant error

Future Value

✷ Future value is defined as present value adjusted for changes in worth

✷ Future value will be greater than present value if the combination of affordability rate and escalation rate indicate that a particular commodity will increase in value over time

✷ It is in essence a weighting of current values to account for future changes in base worth

✷ Future value is not the same as future cost

ANSWERS TO REVIEW QUESTIONS

6.1 Variations in the base worth of goods and services can occur over time due to fluctuations in the standard of living and differential price level changes resulting from product escalation.

6.2 One way is to compare historical records of expenditure on goods and services adjusted for inflation and population growth on an annual basis over time.

6.3 A negative affordability rate means that goods and services are becoming increasingly harder to afford.

6.4 Expenditure and income data must be expressed in real terms, preferable in today's context, per capita.

6.5 Despite a decline in affordability (purchasing power), our expenditure on goods and services (and thus our standard of living) may increase, and the 'gap' be temporarily filled by credit, borrowing and spending of wealth reserves.

6.6 Yes, since averaged national information can be calculated occasionally and used frequently, is readily available, and avoids use of specific financial data which in many cases may be confidential.

6.7 Yes, since affordability can be conceptualised as filtering through all aspects of our society, and in any case public authorities and corporations are but collections of individual taxpayers or shareholders.

6.8 Future value is present value adjusted for changes in worth brought about by affordability or escalation fluctuations over time, and is real in the sense it is in today's terms but is not real in the sense that calculated values are comparative indicators only and do not represent actual transactions.

6.9 Neither interest nor taxation affect the base worth of goods and services, but are included as part of the discounting process.

6.10 Using techniques such as value management or cost-benefit analysis, subjective considerations can be incorporated formally into the decision-making process.

REFERENCES

Flanagan, R & Norman, G (1983), *Life Cycle Costing for Construction*, Surveyors Publications.

McGeorge, D (1988), 'Life cycle costing: What's in a name?', *The Building Economist*, vol. 27, no. 3, pp. 4–5.

Zwartz, S & Marcus, D (1989), 'Living standards', *Consuming Interest*, no. 41, pp.4–9.

CHAPTER 7

REAL AND EQUIVALENT VALUE

LEARNING OBJECTIVES
In this chapter you will learn about the difference between comparison and measurement as applied to life-cost studies. While comparison uses discounted values, measurement activities do not. By the end of this chapter you should be able to:
- understand the concept of discounted future value,
- fully describe the cost/value relationship,
- differentiate between discounting and non-discounting scenarios, and
- explain the components of the composite discount rate and advise on its calculation.

INVESTMENT APPRAISAL AND COST CONTROL

While it is acknowledged that discounting has relevance to life-cost studies, this relevance is confined to situations involving some form of direct comparison. A comparison may consist of just two alternatives, one of which may be described as 'doing nothing'.

The two main objectives of life-cost studies applied to the construction of buildings and their subsequent usage are:

- To facilitate the effective choice between various design solutions for the purpose of arriving at the best value for money. This is a comparison activity.

- To identify the total costs of acquisition and subsequent usage of a given design solution for the purpose of budgeting, planning and controlling actual performance. This is a measurement activity.

Most literature on life-cost studies focuses on the role of discounting because it has been routinely assumed that the technique is about making effective design comparisons. This logic is now refuted.

The investigation of cost and subsequent action is more effective when undertaken at an early stage in the design process. It is therefore not surprising that life-cost studies have concentrated on the first objective, to the virtual exclusion of the second. It is only comparison activities to which discounting relates. Life-cost studies must additionally become concerned with the measurement of costs if value for money is to be rigorously pursued. This involves the presentation of costs in real values rather than comparative (discounted) values.

The measurement of real value involves collection of the total expenditure arising from a project over the study period. Obviously future costs cannot be merely added without some form of adjustment taking place. Inflation is advocated as the appropriate exchange rate for this purpose. Present value (as opposed to discounted present value) translates future cost into real terms, enabling both capital and operating costs to be properly interpreted. Present value thus forms a suitable basis for the measurement and control of life-costs.

The life-cost plan is used to record the present value of both initial and recurrent costs for the chosen design. Determination of the present value of goods and services over the life of a project can enable quantification of the potential cost liability to the owner or investor, which is of great use in the management of financial resources.

Comparative life-cost studies should be undertaken systematically on significant areas of expenditure using a discounting approach. The results of these studies lead to selection of materials and systems and are ultimately reflected in the life-cost plan. Although discounting is a suitable means of assessing the impact of timing on expenditure, the results obtained do not represent real values and there appears little need to present information in this form in the life-cost plan.

Chapter 11 looks at the wider context of making effective decisions based on maximising value as opposed to minimising cost.

Comparative life-cost studies are ultimately a subset of other evaluation techniques that assess the wider implications of value, and in this way differ from measurement-oriented life-cost studies, which form a technique in their own right. Discounting is shown to have a role to play, but it is not appropriate to all the activities to which life-costing is concerned and applied.

CAPITAL BUDGETING VERSUS LIFE-COST STUDIES

The approach adopted for comparative life-cost studies is clearly derivative of DCF techniques employed in capital budgeting. Life-cost studies have been a relatively recent invention. Both are concerned with the adjustment of future cash flows into present value equivalents and with the interpretation of these values for the purpose of ranking and selecting alternate courses of action. Nevertheless, there are several notable differences:

- Capital budgeting concerns the evaluation of investment projects in the context of other available opportunities. Life-costing normally operates at a more detailed level and is concerned with the identification of cost-effective design solutions for a particular investment project.
- It is essential that capital budgeting take both expenditure and revenue into account. The purpose of this approach is to determine the relative productiveness of investments, and this cannot be found if only negative cash flows are considered. Life-costing, which deals only with negative cash flows, is apparently not so constrained.
- Since revenue is not included in life-cost studies it must be assumed when making comparisons that generated income is equal among all alternatives and is thus cancelled out. Differential (offsetting) revenue should be considered but seldom occurs at the detailed level of design choice.
- Competing investment projects may exhibit different expected returns and are selected on the magnitude of those returns. The selection basis for life-costing, however, is minimum total cost.
- The discount rate used in both capital budgeting and life-costing is based on opportunity cost. In capital budgeting profit and risk also affect the decision, and whether included in the discount rate or not are nevertheless involved in the outcome. In life-cost studies profit is automatically maximised as total cost is minimised, and risk is usually assessed by analysing the sensitivity of key variables on the proposed selection.
- The study period used in both capital budgeting and life-costing is usually based on either social welfare or investor interest. In capital budgeting there is a strong tendency to base the study period on economic or useful life during which a positive contribution to society is made, while in life-costing the tendency is towards the period of financial interest of the owner that focuses on the contribution made to the investor.

Profit and risk must be excluded from the discount rate as applied in life-cost studies.

The nature of capital budgeting and comparative life-cost studies is therefore different. Regardless of the previous resolution that the discount rate should be based on the weighted cost of capital, traditional capital budgeting techniques can still provide valid results if profit and risk are additionally included. This is because, even though such matters are not well represented by the compound interest approach, they remain critical to the evaluation of competing investments. But if the discount rate for comparative life-cost studies were to also include profit and risk, incorrect decisions would often result.

DISCOUNTED FUTURE VALUE

The comparison of alternative design solutions is a common activity in the early stages of a project's planning process. Costs and benefits that are not equal among all solutions need to be fully considered in order to arrive at a decision that assists identification of the best value for money. Thus differential revenue, taxation and residual (or sale) value are involved, along with the impact of alternate financing arrangements.

Two or more given alternatives spanning a number of years cannot be realistically compared without matters pertaining to capital productivity and time preference being taken into account. Present value does not consider either aspect. Future value deals specifically with time preference but does not reflect the opportunity cost of money. Hence when comparing alternative solutions the future value of costs and benefits must be undertaken first to determine the equivalent base worth, and then discounted by a rate reflecting the cost of finance considered relevant for the project. The results obtained can be called discounted future value.

Discounted future value merges the philosophies of capital productivity and time preference. Using this approach investment return, inflation, taxation, escalation and affordability are all considered in the discounting process and lead to the assessment of equivalent value. When comparing alternatives in practice, discounted future value can be calculated from present value in a single operation rather than involving future value as an intermediate and often unnecessary step.

A composite discount rate is needed for such purposes, and is determined as follows:

$$d = \frac{(1+i)(1+r)}{1+e} - 1 \qquad \text{(Eqn 18)}$$

where: d = composite discount rate p.a. (factor)
  i = investment return p.a. (%) divided by 100
  r = affordability rate p.a. (%) divided by 100
  e = escalation rate p.a. (%) divided by 100

A suitable approximation of the composite discount rate is $d = i + r - e$. Inflation cancels out yet remains implicit in the process. The discount rate involves prediction of investment return after tax, escalation and affordability into the future. Because discounting is used only for comparison purposes, the inaccuracies of forecast affect each proposal and its impact is largely nullified.

Formulae for discounted future value are identical to Equations 8 and 9, the method of calculation of the discount rate (Equation 18) being the only difference:

$$DFV = PV(1 + d)^{-n} \qquad \text{(Eqn 19)}$$

where: DFV = discounted future value ($)
PV = present value of single payment or receipt ($)
d = discount rate p.a. (factor)
n = number of years

and

$$DFV = \frac{PV[1 - (1 + d)^{-n}]}{d} \qquad \text{(Eqn 20)}$$

where: DFV = discounted future value ($)
PV = present value of annual payment or receipt ($)
d = discount rate p.a. (factor)
n = number of years

For convenience in practice the composite discount rate may be referred to simply as the *discount rate*, and discounted future value may be referred to simply as *discounted value*.

It is therefore recommended that all comparisons calculate costs and benefits in terms of discounted future value. This can be achieved in two ways. Present value can be converted into future value through the determination of the change in worth. The result can then be discounted using a capital productivity discount rate. Alternatively, a composite discount rate can be determined and applied directly to the present value of costs and benefits.

THE COST/VALUE RELATIONSHIP

The proper interpretation of value implies that aspects of cost are converted into present-day equivalents. Although value can be measured only by comparison with other values, its monetary representation can be determined with either real or equivalent figures.

Four types of value, expressible objectively, are identified in terms of their relevance to life-cost studies as follows:

- *Present value* This is today's value of payments or receipts and is equal to the future cost expressed in real terms. It forms the most appropriate basis for the estimation and presentation of life-costs and the interpretation of historical information obtained from life-cost analysis. The present value of individual capital and operating costs, if itemised in the life-cost plan, enables control of actual per-formance to occur, as conversion to future cost can again be made through application of known inflationary indices. As present value does not entail discounting or other worth adjustment, it is a measure of 'real' value that can be used to assess the relative magnitude of different types of life-costs. Intangible costs and benefits where incorporated in a comparison are also measured in present value terms, as they are not consistent with the discounting concept.
- *Future value* This can be described as the theoretical value of goods and services to future generations. Future value measures expected fluctuations in the basic worth of costs and benefits through the consideration of escalation and affordability changes. Life-costs expressed in these terms represent 'equivalent' or com-parative value based on time preference. Future value provides the raw expenditure data for use in traditional capital budgeting and can be calculated without investor-specific matters like financing, taxation and risk assessment being involved.
- *Discounted present value* Commonly (yet undesirably) referred to extensively in the literature as 'present value', this type of value considers the opportunity implicit in the timing of all relevant cash inflows and outflows. It can also be described as 'equivalent' value. As it does not take account of changes in basic worth, dis-counted present value is considered an oversimplification and its suitability to economic appraisal is thus diminished. Building eco-nomists have concentrated too much on the opportunity approach to value determination in the past and have by their actions accepted that time preference is fully subsumed or ignored.
- *Discounted future value* This is considered the correct format for comparison of alternatives. To enable valid economic comparisons to be made among alternatives, all dollar values must be expressed with identical purchasing power reflecting the opportunity cost associated with delayed payment or receipt. Discounted future value combines both the capital productivity and time preference philosophies through adjustment for the time value of money and changes in worth using classical discounting procedures. Discounting results in an abstract representation of value suitable only for comparison and selection of cost-effective solutions.

There is clear advantage in presenting cost information in present value terms in a life-cost plan, even though discounted value is the approach used to justify decisions. The purpose of any cost plan is to set cost targets so that subsequent control and optimisation can occur. Actual expenditure can be de-escalated (using known cost adjustment indices) to the date of the original life-cost plan and related directly to these cost targets. Areas of high expenditure can be investigated and, if necessary, remedial action can be taken. Discounted present or future values do not afford this same facility.

Cash flows expressed in future value terms can simplify later capital budgeting techniques. The increasing use and greater sophistication of life-cost studies in building applications have made it possible and necessary to model or simulate realistically the cash flows throughout the lifetime of a building, and these cash flows over time are the data from which the discounted cash flow techniques used in capital budgeting derive their overall measures of return on investment.

A summary of the cost/value relationship and the terminology used throughout this book is graphically shown in Figure 7.1.

Figure 7.1 The Cost/Value Relationship

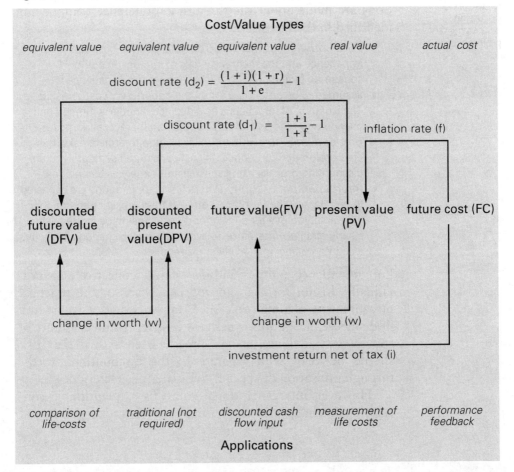

Cost/Value Types

| equivalent value | equivalent value | equivalent value | real value | actual cost |

$$\text{discount rate } (d_2) = \frac{(1+i)(1+r)}{1+e} - 1$$

$$\text{discount rate } (d_1) = \frac{1+i}{1+f} - 1 \qquad \text{inflation rate (f)}$$

| discounted future value (DFV) | discounted present value(DPV) | future value(FV) | present value (PV) | future cost (FC) |

change in worth (w)   change in worth (w)

investment return net of tax (i)

| comparison of life-costs | traditional (not required) | discounted cash flow input | measurement of life costs | performance feedback |

Applications

It is not practical to discount costs and benefits by a multitude of different rates that reflect changes in their individual base worth. This must often be a separate process and, due to the detailed considerations involved, a specialised one.

A COMPOSITE DISCOUNTING PHILOSOPHY

Although there are two philosophies for the determination of the discount rate, most literature recommends capital productivity as the preferred approach. Certainly a discount rate based solely on one or more of the time preference components (pure impatience, risk and uncertainty or intergenerational equity) would require explicit consideration of interest received on account balances and interest paid on outstanding loans. Perhaps the only way to arrive at a discount rate based on fundamentally myopic criteria is to use opportunity cost as a surrogate.

Refer to Chapter 3 for the previous discussion concerning time preference.

In order to clarify this discussion, the following observations are made about the three time preference components identified in the literature:

- *Pure impatience* There is an obvious overlap between impatience and the theoretical loss of interest on equity consumed by the project, and for this reason impatience is assumed to be accounted for by the opportunity cost principle. The isolation of impatience as a distinct variable is difficult and may be regarded as non-observable for all practical purposes.
- *Risk and uncertainty* In a similar fashion to project risks, this time preference component is best considered separately to the discounting process using specialist risk analysis techniques.
- *Intergenerational equity* Where future generations are expected to be relatively better off or worse off than present generations, their ability to pay for and support decisions made in the past is clearly affected. Therefore, to account for such changes in equity, a counterbalancing mechanism is required.

Pure impatience and risk and uncertainty collectively or individually do not form an appropriate basis for the practical measurement of time preference. Impatience is a vague and ill-defined concept and is perhaps best represented as part of some other attribute. It has already been shown that risk should be dealt with separate to the discounting process through utilisation of specialist risk analysis techniques.

However, intergenerational equity is a quantifiable concept, since changes in the base worth of traded goods and services between present and future generations can be used to indicate relative prosperity. Such changes in worth influence equivalent value used in an economic appraisal. If goods and services are worth more in real terms at some point in the future compared with their worth today, and the level of real income has not matched or bettered this rise, future generations will find such goods and services more difficult to afford. Therefore the costs and benefits estimated in the analysis, which are a function of individual goods and services, need to be adjusted so that they are worth more the further away from the present they occur. Alternatively, the discount rate can be reduced so that it gives more importance to future costs and benefits.

These objections to time preference are discussed in further depth in Markandya and Pearce (1988).

There are, nevertheless, objections to the use of time preference concepts in the discounting process. First, individual time preference is not necessarily consistent with individual lifetime welfare maximisation. Second, individual wants carry no necessary policy implications. Third, a society that elevates want-satisfaction to high status should recognise that it is the satisfaction of wants as they arise that matters, not today's assessment of those wants. And fourth, if the risk of death argument is employed, it is illegitimate to derive implications for immortal societies from the risks faced by mortal individuals. While these objections relate to pure impatience and risk, they do not have an influence on the concept of intergenerational equity.

Intergenerational equity is therefore at the heart of the time preference approach. Where no time preference changes are expected and where future generations are not to be disadvantaged by present investment decisions, then a discount rate of zero appears the only solution. However, finance costs still need to be considered when appraising an investment and, if represented as a discount rate, no bias against future generations is implied. In other words, a discount rate based on the weighted cost of capital supports intergenerational equity principles, because if finance costs were included explicitly in the cash flows then a zero discount rate would result.

Intergenerational equity change is not part of a discount rate based on the after-tax weighted average of equity and borrowed capital, nor is it included in any of the other reviewed variations of the capital productivity philosophy. Therefore this time preference component is being ignored in the appraisal of investment projects, yet future generations will inherit these projects and the responsibility for their management.

The discounting process should compensate for movements in equity between generations where such changes are anticipated. This can be achieved by adjusting the present value of cash inflows and outflows for changes in their base worth over time. A discount rate based on capital productivity would then be determined and applied in order to calculate equivalent value.

CONCEPTUAL FRAMEWORK

A conceptual framework depicting the role and basis of discounting in life-cost studies is shown in Figure 7.2. It divides life-cost studies into its two constituent purposes of measurement and comparison; the latter includes consideration of differential revenue where appropriate. All but the tangible costs and benefits arising from a comparison are expressed in terms of real value. Intangible costs and benefits, such as those resulting from social and environmental effects, are estimated and converted to an annual sum that can be distributed over the selected period of financial interest.

Figure 7.2 Discounting Conceptual Framework

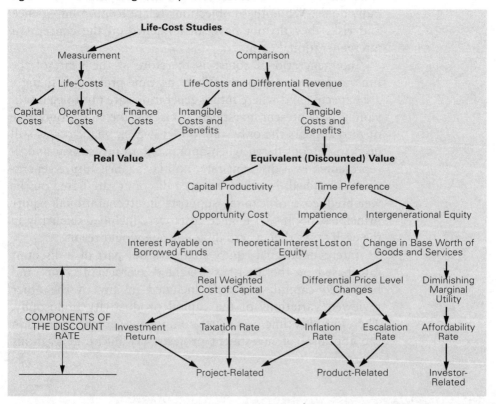

The comparative value of tangible costs and benefits is a function of both capital productivity and time preference. Capital productivity uses an opportunity cost approach to take account of the actual interest payable on borrowed funds necessary for the project to be developed and operated, which is of course a real cash flow, and the theoretical interest lost on the use of equity. The latter is necessary to enable consumption of accumulated reserves to be recognised and additionally forms a useful method for translating impatience and human preference for control over resources into measurable terms. Interest lost and interest payable are weighted according to their likely incidence and are expressed in real terms after tax.

Time preference also encompasses intergenerational equity concerns, which are measured by changes in the base worth of goods and services between present and future generations. Differential price level changes and diminishing marginal utility collectively depict this change. Costs and benefits are held to be represented by the value of traded goods and services.

The discount rate is therefore a function of the real weighted cost of capital, differential price level changes and diminishing marginal utility. Any or all of these matters may be positive or negative, and so the composite discount rate can similarly be of either sign. Investment return, taxation and inflation are variables in the calculation of the real weighted cost of capital. Inflation and escalation are variables in the calculation of differential price level changes. Affordability is promoted as an appropriate variable in the calculation of diminishing marginal utility. When these variables are amalgamated, inflation cancels out but remains implicit in the process.

Profit and risk are two notable exclusions from the discounting process previously described and are best dealt with using a different strategy. Both these factors are of vital importance, but are not well represented by the compound interest principle and when loaded into the discount rate result in undesirable distortions in the calculation of equivalent value.

The discounting conceptual framework embodies a number of recommendations that either in whole or in part further the existing state of knowledge relating to discounting. The recommendations are:

The determination of the discount rate is more fully explored in Langston (1994).

- Discounting does not have universal application to life-cost studies but is confined to activities that involve the comparison of tangible costs and benefits.

- Intangible costs and benefits, such as theoretical environmental and social considerations, are not discounted but should alternatively be expressed as real value.
- The time horizon for life-cost studies is the period of financial interest of the owner or investor. Social and environmental impacts resulting from the project's existence should be annualised across the selected time horizon.
- The rationale for discounting is shown to be opportunity costs, impatience and intergenerational equity.
- Intergenerational equity is measured by the change in base worth of goods and services as depicted by differential price level changes and diminishing marginal utility.
- Diminishing marginal utility can be quantified by analysis of income and expenditure patterns using an affordability approach.
- The composition of the discount rate includes affordability as a variable.
- The discount rate is shown to comprise project-related, product-related and investor-related attributes that collectively may result in a number of discount rates for application in a single comparative study.

IMPLICATIONS FOR LIFE-COSTING

The discounting process has implications for all life-cost studies involving comparison of alternatives. However, planning and control activities are also of significant importance. Life-cost studies can be concluded as comprising three main areas:

- *Analysis* The recording of historical data as an aid in planning and evaluation requires a structured classification system. Life-costs need to be expressed in terms of future cost and indexed by time period.
- *Planning* The measurement and presentation of total costs in terms of real value is essential to demonstrate the true magnitude of different types of life-costs resulting from cost-effective design decisions. Life-costs thus need to be expressed in terms of present value, so as to support control and act as a target against which actual performance can be judged. A yearly cash flow expressed in terms of future value is a useful source of data for subsequent DCF analysis utilising a capital productivity discount rate.
- *Evaluation* Value for money can be attained by comparison of competing design solutions, taking account of taxation, residual value, affordability and the timing of cash inflows and outflows. Tangible life-costs need to be expressed in terms of discounted future value and intangible life-costs in terms of present value. Comparison of alternat-ives can occur prior to or subsequent to construction, but is most effective in the early stages of the design process. Design evaluation is the only part of life-cost studies to which discounting applies.

The past use of discounting is challenged as having an absolute role in life-cost studies. Planning and control are measurement-based tasks and are vital if resources are to be properly managed. Furthermore, acknowledged problems relating to the prediction of economic performance well into the future are largely overcome by documenting life-costs in terms of real rather than equivalent value. The use of the life-cost plan as a control mechanism for future expenditure is thus facilitated and the likelihood of life-cost studies being undertaken in the first place is greatly improved.

Past difficulties with life-costing no longer need to be recounted as barriers, and the challenge before us is to start using the technique effectively in practice. How this is done is discussed in Chapter 8.

Difficulties with traditional discounting-based tasks are often highlighted as the primary reason for the lack of implementation of the total cost approach in practice. The measurement of real value solves many of these difficulties and leads to the following conclusions:

- *Forecasting* The removal of interest, taxation and inflation from the measurement process overcomes previous problems with prediction and avoids much of the criticism over the wide range of solutions that could be produced. While it is recommended that affordability changes be additionally included when undertaking comparisons of design alternatives, errors of judgement in this context affect all solutions and their impact is thus somewhat diminished.
- *Historical data* The need for vast amounts of historical data is largely obviated by calculation of life-costs and residual values from first principles in a similar manner to the estimation of capital costs. Information that must be determined or obtained centres around the lives of the various components within the building and the extent of maintenance work expected. Standard elements and sub-elements form an excellent classification system for storage of both estimated and collected data.
- *Future occurrences* Future decisions to extend, renovate, refurbish or replace, accelerated environmental damage, owner commitment to maintenance and the availability of funds for such work cannot be predicted prior to construction. Fortunately this is not important, as the life-cost plan is merely a target within which future control can then operate. Planning should not be discarded simply because future events may occur to change that plan. Comparison between alternat-ives will assume the same theoretical conditions, and hence valid decisions of construction, materials and systems can still be made.
- *Ownership* It may be true that developers are not interested in life-costs if they plan to sell the building on completion to another party. However, the new owner should be concerned with the likely running costs and may wish to undertake a study prior to purchase. Thus life-cost plans in many cases may be a marketing advantage. Government authorities, which are accountable to the general public, must logically consider life-costs for all projects. Nevertheless, life-cost plans at year zero will have the effect of limiting the study to capital costs. Use of the period of financial

interest of the owner as the basis for the time horizon lends support to this notion.
- Ease of use Although the calculations involved in preparing a life-cost plan are complex and numerous, the application of computers offers substantial benefits. In addition, the preparation of rates per unit for various types of construction and for various lives will make life-costing as simple as capital cost estimating.
- Market valuation The ability to calculate the remaining life (and hence residual value) of each component within the building will assist in the assessment of market value for the property where information on comparative sales is unavailable or unreliable.

REVIEW QUESTIONS

7.1 Why is discounting an abstract measure?

7.2 Is discounting socially justifiable?

7.3 Why is the choice of discount rate one of the most critical decisions in the comparison process?

7.4 Do high discount rates encourage minimum standards of construction?

7.5 What are some of the key differences between discounted cash flow analysis and comparative life-cost studies?

7.6 What are the benefits of measuring life-costs in terms of present value?

7.7 How is revenue incorporated into life-cost comparisons?

7.8 Why is inflation excluded from the equation for composite discount rate (see Equation 19)?

7.9 Why do we express expenditure in the life-cost plan in terms of present value rather than discounted future value?

7.10 Under what situations should we employ discounting?

TUTORIAL EXERCISE

Refer to the tutorial exercise from Chapter 6. Your task this chapter is to prepare a cash flow summary comprising totals for PV, FV and DFV for each year of the nominated 10-year time horizon.

In addition to previously specified economic data, you should also assume that the rate of investment return is expected at 10% per annum. Taxation issues can be ignored in this exercise.

> **HINT**
> Discounted future value can be determined directly by discounting a cash flow comprising future value expenditure.

KEY POINTS

A PowerPoint presentation dealing with the topics discussed in this chapter can be downloaded from the publisher's website (see the publisher's details at the beginning of the book for the address). Some key points are shown below.

Discounting Mystic

* Discounting is nothing more than a comparison tool
* It disadvantages expenditure or revenue in the future against the present to reflect that control over money is beneficial
* Discounting is socially unfair only if the rate of discount is different from the true time value of money
* Discounting accounts for all interest costs

Components of the Discount Rate

* The discount rate is a function of investment return, inflation, taxation and change in worth
* Profit and risk should be excluded from the discounting process and assessed in other ways
* The discount rate is a reflection of the true time value of money
* Discount rates are a combination of project- based, product-based and investor-based considerations

DCF Differences

* Discounted cash flow analysis and life-cost studies differ in a number of significant ways

* DCF requires both expenditure and revenue to be taken into account, whereas life-costing is expenditure-based

* DCF selection criteria concern maximum return whereas life-costing is about minimum cost

* DCF focuses on economic life whereas life-costing is related more to holding period

Life-Cost Applications

* Life-cost studies can be divided into two separate activities of comparison and measurement

* Comparison facilitates the effective choice between various methods of achieving a given objective for the purpose of arriving at the best value for money

* Measurement identifies the total costs of acquisition and operation as a means of control

* Both activities are important

Comparison

* Comparisons between two or more alternatives require the use of discounting

* Costs are reduced according to the time period in which they are incurred so that expenditure that is deferred is given less weight

* Comparative studies are important because they can identify cost-effective solutions

* Alternatives are expressed in equivalent value terms as a means of selection

Measurement

* Measurement relates to the specific study of a given design
* Cost plans and cash flows are typical measurement outcomes
* Discounting is not involved
* Costs are expressed in real terms (i.e. present value)
* This aspect of life-cost studies has traditionally been given little attention

Types of Values/Costs

* Present value should be used for all measurement activities
* Future value is similar to present value but includes changes in worth
* Discounted present value is normally used in practice for comparative studies, but is now subsumed by discounted future value
* Future cost is the actual 'cheque-book' amount but cannot be added across time periods

ANSWERS TO REVIEW QUESTIONS

7.1   Discounting is an artificial mechanism invented so as to disadvantage future events over the present in a similar way to that in which we value money today over money tomorrow, but discounted values do not represent real money and are used solely to assist us in assessing complex cash flow patterns with objectivity.

7.2   Yes, provided the discount rate is based on the true time value of money and is not loaded so as to bias present generations over future generations, since this rate then merely reflects the interest costs associated with financing the project and is therefore a normal part of the procurement process.

7.3   The discount rate embodies all the economic predictions (such as inflation, interest rates, escalation, affordability, taxation and equity proportion) into a single variable, and its impact over long time frames can have a significant effect on the outcome of the analysis.

7.4   Yes, since high discount rates will rapidly diminish future cash flows relative to initial construction costs and therefore favour projects that can minimise initial costs usually at the expense of subsequent operating performance.

7.5   The three main differences are that life-costing considers costs (not revenue), bases selection on minimum cost (not maximum return), and focuses on the period of financial interest (not economic life).

7.6   Present value enables interpretation of life-cost studies today, avoids predictions based on economic factors, can be adjusted to future time periods through the use of known inflationary indices, and enables assessment of the relative magnitude of different types of life-costs.

7.7   The difference between revenue flows for alternate design solutions can be treated as a negative cost in much the same way as residual value and tax concessions are handled, although normally differential revenue cannot be easily estimated unless total projects are compared.

7.8   As this formula is the combination of two other formulae (Equations 7 and 16), general inflation (f) is cancelled out, although specific inflation (e) still remains.

7.9   Life-cost plans are measurements of specific projects and do not involve comparisons with other projects, so discounted values are not relevant and would mislead the client into thinking that the discounted values were real.

7.10   Discounting is correctly applied to comparative situations involving cash flows over a number of years, and is used to calculate equivalent values so that income and expenditure in different time periods can be properly evaluated.

REFERENCES

Langston, C (1994), 'The determination of equivalent value in life-cost studies: An inter-generational approach' (PhD Thesis), University of Technology, Sydney.

Markandya, A & Pearce, DW (1988), 'Environmental considerations and the choice of discount rate in developing countries' (Environmental Department Working Paper No. 3), World Bank, Washington.

CHAPTER 8

LIFE-COSTING IN PRACTICE

LEARNING OBJECTIVES
In this chapter you will learn how to implement life-cost studies in practice and what other techniques and procedures are required. Computer software can greatly assist in ensuring that calculations are correct and presentation formats are consistent. By the end of this chapter you should be able to:
- describe the context within which life-cost studies operate,
- outline the five main stages of project cost control,
- discuss the strategy for wider implementation of life-cost studies into practice in the future, and
- be familiar with the type of information typically produced.

NEW CHALLENGES FOR LIFE-COST STUDIES

It is necessary that a methodology be developed that will encourage the adoption of the total cost approach by building economists operating within the global building industry, and thereby contribute to the attainment of value for money to investors. There is still a lot of progress to be made in this field, especially when it comes to getting methods used in practice.

Much debate has been raised over the years about the advantages and disadvantages of taking the total capital and running costs of a building into account in order to ultimately assess value for money. Although the disadvantages are numerous, the fundamental basis for interpreting future costs has seldom been included among them.

'The way ahead for life cycle costing is likely to be greater attention being paid to the cost/value relationship. ...The development of life cycle costing techniques to take account of values is a very short step to take, but the complexity of the issues should not be underestimated. Whilst much work has been undertaken to bridge the gap between the theory of life cycle costing and the potential realities, there is still a considerable amount of development work to be completed.' (Flanagan 1985, p. 15)

Hollander (1976, p. 20) states that the technique has its roots in engineering economics, 'is very well known in academic and engineering circles, and the application of this technique has existed for about a century'. Picken (1987, p. 46) elaborates by observing that 'certainly the economic and mathematical techniques have been known for a long time, for example Wellington's "The Economic Theory of Railway Location" published in 1887'. However, it was not until 1960 that building economists recognised the potential of the total cost approach.

Past concentration on discounting the costs of competing design solutions has been the major impediment to the widespread use of life-costing techniques, as it is this aspect that encourages and supports most criticism.

To overcome this barrier it is necessary to view the total cost approach in its broadest sense and to recognise that the comparison of alternatives, to which discounting applies, is only one activity within this larger discipline.

Discounted cash flow analysis has been advocated as the technique to be used for comparing future dollars with their equivalent value today. It has long been employed with success in the assessment of investment opportunities that possess differential timing in the payment and receipt of cash, but its specific utilisation in life-cost studies has been more recent.

The application of comparative life-cost techniques is likely to be fruitless unless accompanied by other control activities. Life-costing can also be applied to assist in the logical management of an asset without looking at alternatives. Benefits of this approach include identifying cost-significant areas, establishing funding requirements and timing, setting targets for expenditure control and obtaining a clearer understanding of total cost commitment. The facility management process is now beginning to recognise the importance of life-cost studies to the management of existing facilities.

In times of rapidly rising prices owners will seek every avenue to improve value and conserve resources. While comparison of alternative design solutions is fundamental to obtaining value for money, the measurement of life-costs in real terms also possesses significant merit.

Capital cost control has long been understood to comprise numerous activities, including but not limited to the comparison of alternatives. In many countries there is an obvious bias towards the measurement and costing of a given design, due principally to the architect's dominance within the professional design team and the frequent engagement of the building economist after concept formulation has been completed. Perhaps life-costing should initially concentrate more on measurement and control activities.

Some investors are becoming interested in knowing more about what performance can be expected from their buildings and in considering costs over the period of their interest in the building project as part of understanding the worth of their capital investment. They identify this trend as indicative of the new challenge for life-cost techniques. Furthermore, advice on maintenance, replacement and other operating costs are now seen by investors as areas of unsatisfied demand. Commonly, design decisions are made without reference to operating costs.

Past failure to obtain widespread practical acceptance for the total cost approach is largely due to the existence of several key drawbacks that have achieved frequent and continued publicity. While some cite the inability to predict future trends and events with any measure of certainty as the principal drawback, others lay the blame variously at the doors of capital cost controls, fragmentation of responsibilities for the finance of capital and revenue expenditure, difficulties of obtaining sufficient and appropriate historic data, and of agreeing the basis for calculation.

Certainly forecasting is unavoidable with any technique that aims to improve on a first-cost approach. Forecasts can be seen as more reliable, however, if economic predictions, such as inflation and investment return, are employed only in comparative studies where the effect of their likely inaccuracy is minimised. Historical data can help forecasts of maintenance periods and operating costs to be realistic, especially where the data source is the actual project for which it will be reused. A systematic approach for life-cost analysis can substantiate the data employed during life-cost planning activities, and over time provide a degree of confidence that will dispel criticisms of inaccuracy.

It remains the case that many organisations have impenetrable barriers in place between capital and revenue budgets. This mentality seriously undermines implementation of a total cost approach. Public authorities are principal offenders. Revenue is often considered as a function of the level of expenditure necessary to maintain and operate the building.

Taxation also plays an important role. Building work that is depreciable at enhanced rates and has operating costs that are fully deductible is normally viewed with favour. For example, an expensive air conditioning system might be selected over the alternative of sunshading and natural ventilation due principally to taxation concessions.

Any new ideas must, if they are to have a direct influence on how the industry operates, be capable of implementation. The importance of this statement should not be underestimated in any new initiatives. Life-cost research has generally concentrated on theoretical aspects at the expense of practical solutions.

'The logical step is to extend our skills and to bring within our control the outstanding balance of the total cost which an owner will face and present to him a complete financial picture for the whole economic life of his building. In this way we can provide him with a building that is not only designed to cost but which can be planned and built to operate to cost as well.' (Durden 1983, pp. 174–175)

'The building industry needs technically correct, but practical, methods and guidelines for evaluating the economic performance of alternative building technologies in a consistent manner.' (Marshall 1987, p. 23)

THE COST CONTROL PROCESS

If asset management is the global term covering all aspects of acquisition and utilisation of fixed assets, then capital budgeting and cost control are clearly two of the main supporting processes. Capital budgeting, comprising techniques such as discounted cash flow and cost-benefit analysis, enables fundamental decisions to be reached about the type and extent of the investment. Profit maximisation is a key objective. The cost control process, on the other hand, begins with the decision to acquire the asset and functions until its cost implications are no longer relevant to the owner. Continual interaction between the capital budgeting and cost control processes commonly occurs. The objective of cost control is the attainment of value for money for a particular design solution, and it is to this process in particular that life-cost studies have clear application.

Figure 8.1 illustrates an implementation framework for the cost control process applied to building projects. At the heart of the process are the activities of life-cost planning and life-cost analysis. An elemental approach is preferred as the basis for cost estimation and data collection activities throughout the process.

Strategic planning, design and construction and/or facility management form the three main processes in the framework. Life-costing is concerned with the latter two, identified as dealing with project cost control objectives.

Strategic planning is not dealt with specifically in this book. A good reference is Best & de Valence (1999).

The cost control process itself can be divided into five stages. These stages depict procedural milestones rather than a chronological sequence. They draw on other techniques such as cost modelling, value management and post-occupancy evaluation in order to maintain control and deliver value for money to the investor. The identified stages and their corresponding documentation requirements are:

- *Client brief* This stage draws on historical data in order to formulate a budget representative of the investor's needs, which can be used thereafter as a comparison tool between actual and expected performance. The budget can be considered as an upper limit for funding purposes, with the success of the cost control process measurable by the savings achieved.
- *Optimisation* This is the most important stage in the cost control process. Design options are proposed to meet the investor's needs and are subsequently evaluated. Using value management techniques incorporating life-cost comparisons, a

schematic design is developed and concept sketches prepared. The nature of the design and the selected combinations of materials and systems are reflected in the sketch design life-cost plan and their cost reconciled against the budget.

- *Documentation* Final working drawings are prepared in association with the specification. Cost checks ensure that the developing design continues on track. The tender document life-cost plan details the completed design and acts as a management tool for the remainder of the building's life. The life-cost plan can alternatively be produced from a summary of the bill of quantities where one exists.
- *Monitoring* Actual performance during construction and/or occupation of the building is monitored for two purposes: first as an input for ongoing management and second as a feedback source for future budget preparation. Activity reports are regularly prepared, and include progress reports, expenditure reports, maintenance requirements and component lives. Post-occupancy evaluation also occurs during this stage and assists in the preparation of future value management studies.
- *Management* This stage comprises actions that management may take, in accordance with policy decisions, to improve the existing design. Such recommendations are listed in occasional management action plans. Value management techniques also have an important application here.

Figure 8.1 Project Implementation Framework

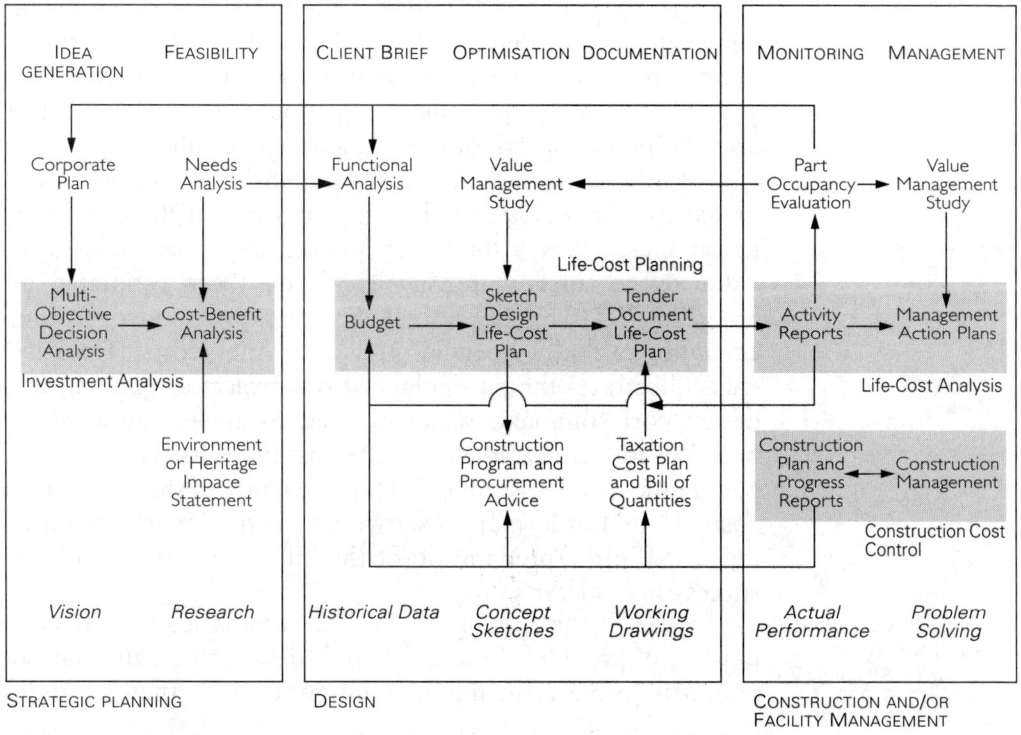

The first three stages of the cost control process constitute life-cost planning and are accordingly part of the design process. The latter two stages constitute life-cost analysis and apply equally to both construction and facility management processes.

CASE STUDY

Albion Park High School, designed to accommodate a thousand secondary school students, has been built at Albion Park on the south coast of New South Wales, Australia. Its construction was supervised by the New South Wales Public Works Department on behalf of the Department of School Education. The design comprises numerous identifiable functional areas physically subdivided into blocks: namely, (1) Gymnasium and F.S.U., (2) Administration, (3) Performing Arts, Music and Home Economics, (4) General Learning, (5) Science, (6) Industrial Arts, (7) Arts, (8) Library, and (9) Agricultural Science. The project was completed and in service early in 1991.

The complete case study was used as a pilot to demonstrate the benefits of planning and analysing building performance over time. Albion Park High School is the first Public Works Department project to have its operating costs annually monitored and reconciled against the targets set down in the tender document life-cost plan. While not all projects are expected to be regularly monitored, data can be collected and applied to new designs from a selection of past projects that are considered typical for their functional classification.

The life-costs for Albion Park High School have been estimated over a period of financial interest of 100 years. The target life-cost over this study period, expressed in present value terms current at May 1989, has been estimated at $36 745 953. This figure includes all capital and operating costs but excludes financial and occupancy costs. The case study illustrates the particulars of the project as given in the design cost summary, which is used as an 'executive summary' of the tender document life-cost plan. Its purpose is to communicate the overall results of the study to the client. The costs relate to information shown on the final working drawings used in the preparation of the bill of quantities and the successful tenderer's bid.

The pages described for this study are shown in Appendix 2 (pages 217–228).

Page 1 is the title sheet. Page 2 summaries the life-cost plan and provides locational and descriptive information about the project, including assumptions and efficiency indicators. Pages 3-6 contain the accumulated elemental cost summaries for all blocks and external works. Pages 7-8 summarise operating costs in the categories of cleaning, energy, repair, replacement and 'other' by element. Page 9 provides expected cash flow details over the life of the project.

Page 10 shows the distribution of life-cost by type. Clearly capital costs are the most significant (30.04%), followed by replacement costs (26.98%), energy (15.85%), repair (11.15%), cleaning (8.72%) and other costs (7.26%). Land purchase, selling costs, council rates and finance costs are not applicable, while occupancy costs are excluded. Page 11 divides these life-costs into the major elemental groupings. Pages 12 and 13 show capital and operating costs (respectively) by function. Similar proportions of capital and operating expenditure were discovered. Operating costs can also be collected by year to produce a cash flow useful for forward planning and control. Page 14 illustrates the annual cash flow in terms of present value.

See Appendix 2 on pages 217–228 for details of the study.

This case study is one of a few examples of a life-cost investigation that divorces itself from the shackles of discounting, instead preferring to express all costs in terms of real values. The information format is meaningful to the investor and the project consultants and identifies elemental targets suitable for effective cost management. The case study therefore provides evidence that discounting and the determination of equivalent value are not all-pervasive in life-cost studies, a point that is generally overlooked in the literature.

The case study can be found in full in LIFECOST and comprises a budget, sketch design life-cost plan, tender document life-cost plan, design cost summary and examples of a cost check, activity report, management action plan and value management study. The capital cost and quantity information used in the case study are derived from the cost plan and bill of quantities prepared by the consultant quantity surveyor. The operating cost information is estimated generally from first principles, although some historical data were available. The information presented uses concepts that have been discussed in the previous chapters.

LIFECOST is a Microsoft Excel template developed originally by Computerelation Australia Pty Limited and available via the author of this book as a teaching tool. It is sold under an educational site licence.

REVIEW QUESTIONS

8.1   How can the separation of capital and revenue budgets be discouraged?

8.2   What is the link between life-cost planning and life-cost analysis?

8.3   Why is the optimisation stage of cost control the most important?

8.4   Life-costs need to be expressed in different terms for different purposes, so what are the 'units' for life-cost planning, life-cost analysis and design evaluation?

8.5   How can the uncertainty of economic prediction be diminished when undertaking life-cost comparisons?

8.6   How can a bill of quantities assist in the preparation of a tender document life-cost plan?

8.7   Are there still unsolved issues associated with life-costing that will inhibit its effective use in practice?

8.8   What is a 'design cost summary' and when is it required?

8.9   What are the five main stages of the cost management process for life-cost planning and analysis?

8.10   What are the advantages of using software to undertake life-cost investigations?

TUTORIAL EXERCISE

The application of life-cost studies to the construction industry can be extended to include understanding of the relationship between cost and energy. Research published by Langston and Ding (2004) has determined the data shown below.

Each project is an Australian high school. Energy includes embodied energy (based on an analysis of the abbreviated quantities shown in the life-cost plan) and operating energy (based on actual usage plus estimated recurrent embodied energy).

The following data are based on a 60-year time horizon. Determine the relationship between total life-cost ($/m^2) and total energy consumption (GJ/m^2) using regression analysis. What is the strength (i.e. reliability) of this relationship?

Project	Floor Area (m2)	Total Life-Cost (A$2002)	Total Energy (GJ)
1	9 800	30 596 967	416 904
2	13 519	35 983 137	457 260
3	8 092	23 884 337	367 354
4	12 565	40 120 613	470 037
5	11 398	30 811 079	377 819
6	15 344	43 041 059	518 274
7	5 268	20 308 581	247 410
8	8 610	27 215 715	356 731
9	12 265	35 936 535	420 474
10	10 747	35 077 732	529 469
11	13 213	39 815 840	450 205
12	3 040	11 817 083	341 755
13	9 220	28 475 273	341 629
14	1 295	6 038 010	114 923
15	8 864	29 861 059	364 856
16	8 500	27 717 456	380 930
17	15 631	44 957 713	649 612
18	4 066	20 697 792	435 216
19	7 516	22 183 221	357 830
20	3 524	11 502 302	92 778

HINT
Use linear regression.

KEY POINTS

A PowerPoint presentation dealing with the topics discussed in this chapter can be downloaded from the publisher's website (see the publisher's details at the beginning of the book for the address). Some key points are shown below.

Conceptual Framework

❋ Life-cost planning and life-cost analysis can be viewed in the context of a conceptual framework to effectively illustrate the linking of typical activities

❋ Five stages are used, comprising client brief, optimisation, documentation, monitoring and management

❋ The first three stages concern life-cost planning

❋ The last two stages concern life-cost analysis

Client Brief

* The objective of this stage is to interpret the client's requirements, to set elemental targets and to establish overall cost limits for the project

* Historical data form the information source employed

* The main document produced during this stage is a budget

* This is a prime cost control document and forms a reference point for the processes that follow

Optimisation

* The objective of this stage is to investigate alternatives that satisfy the client's requirements and to determine which alternative represents the best value for money

* Concept sketches are the primary information source

* The main document produced at this stage is a sketch design life–cost plan

* Value management studies are useful inputs

Documentation

* The objective of this stage is to provide detailed information about the developed design that will be used as a planning tool throughout the remaining life of the project

* Working drawings are the primary information source

* The main document produced at this stage is a tender document life–cost plan

* Cost checks are prepared between cost plans

Monitoring

✳ The objective of this stage is to collect information pertaining to the construction and occupation phases of the project as a feedback source for future projects and as an input to ongoing management decisions

✳ Actual performance is the primary information source

✳ The main documents produced at this stage are regular activity reports

Management

✳ The objective of this stage is to make recommendations and implement improvements pertaining to the construction and occupation phases of the project

✳ Revised details are the primary information source

✳ The main documents produced at this stage are occasional management action plans

✳ Value management studies are useful inputs

LIFECOST Software

✳ The LIFECOST software has online help containing many pages of guidelines and definitions for life–cost planning and analysis

✳ It also comprises over 300 pages of fully worked examples for each of the document types previously described

✳ The worked examples in Appendix 2 relate to a real project called Albion Park High School built on the south coast of New South Wales (Australia)

ANSWERS TO REVIEW QUESTIONS

8.1 Link asset management processes involved in strategic planning, design, construction and facility management to enable a more holistic approach to cost management, and enlighten the client regarding the benefits of spending more initially to deliver overall savings.

8.2 Cost control is a part of total asset management and follows the investment analysis (strategic planning) process beginning with the decision to acquire an asset and functioning until its cost implications are no longer relevant to the owner.

8.3 Life-cost analysis provides feedback for life-cost planning through the monitoring and collection of actual performance so that future design decisions can be improved.

8.4 It is generally accepted that the greatest opportunity for saving occurs at the early stages of design before time and effort is wasted documenting solutions that are inherently cost expensive.

8.5 Life-cost planning is expressed in present value terms, life-cost analysis in future cost terms indexed by time period, and design evaluation in discounted future value terms.

8.6 The bill of quantities, if coded according to elemental classifications, can be compiled into a tender document format so that the time normally required to prepare the cost plan is significantly diminished.

8.7 Theoretical and conceptual issues have now been overcome, but a case still needs to be made to convince clients that spending money on fees for this service will realise additional benefits, and further education of practitioners is required.

8.8 Every tender document life-cost plan is to be accompanied by a design cost summary, bound separately, to facilitate the effective communication of the project's cost implications to the client.

8.9 The main stages are client brief, optimisation, documentation, monitoring and management.

8.10 Speed, accuracy, consistency of presentation and categorisation, as well as possible links to add models, are the primary advantages of computer software tools.

REFERENCES

Best, R and de Valence, G (1999), *Building in Value: Pre-design Issues*, Butterworth-Heinemann.

Durden, M (1983), 'Recurrent costs: The increasing importance', *The Building Economist*, vol. 22, no. 4, pp. 174–176.

Flanagan, R (1985), 'Life cycle costing: What has happened in the United Kingdom' in proceedings of International Comparison of Project Management and Performance Conference, Melbourne.

Hollander, GM (1976), 'Life-cycle cost: A concept in need of understanding', *Professional Engineer* (Washington), vol. 46, no. 6, pp. 20–22.

Marshall, HE (1987), 'Survey of selected methods of economic evaluation for building decisions' in proceedings of Fourth International CIB Symposium on Building Economics, Copenhagen, pp. 23–57 (keynotes).Langston, C and Ding, G (2004), 'Multiple criteria sustainability modelling: A case study', *International Journal of Construction Management* (Special AUBEA Edition), in press.

Picken, DH (1987), 'Life cycle costing: Can it be effective?' in proceedings of Fourth International CIB Symposium on Building Economics, Copenhagen, pp. 43–51.

CHAPTER 9

OCCUPANCY COSTS

LEARNING OBJECTIVES
In this chapter you will learn that the typical costs considered during the design of buildings are but a small component of total costs when occupancy is considered. The costs of functional operation are dominated by staffing expenditure. By the end of this chapter you should be able to:
- identify types of occupancy costs,
- understand strategies that can be used during project design to minimise future occupancy costs,
- comment on possible productivity gains, and
- relate some success stories from past projects.

WHAT ARE OCCUPANCY COSTS?

Occupancy costs are one type of operating cost and hence are a component of a building's life-cost. Occupancy costs include staffing, manufacturing, management, supplies and the like that relate to a building's functional use. The term *functional-use cost* is sometimes used as an alternative descriptor. Occupancy costs also include denial-of-use costs caused by delays in the acquisition of the facility due to construction, renovation and the like.

Occupancy costs can be subdivided into the following categories:

- *Acquisition expenditure* This comprises outlays for initial purchase of goods and services related to the functional use of the building. Examples include office fit-out, stationery, computers, medical supplies, etc. Denial-of-use costs are normally included here.
- *Recurrent expenditure* This comprises the ongoing costs of maintaining the functional use of the building. Examples include staff salaries, manufacturing raw materials, washing and sterilising expenses, replacement office supplies, etc.

A study of a CBD building in Sydney (Australia) over a 50 year life revealed that 11% of the life-cost was capital, 82% was operating and 7% was finance costs. Of the 82%, 18% was cleaning, energy, repair, replacement and other miscellaneous expenses and 82% was occupancy costs. Therefore 67% of the life-cost of the building over 50 years was related to its functional use. Often this proportion can be substantially higher, particularly in the health care and manufacturing sectors.

It is inappropriate to consider revenue here since occupancy cost is a subset of life-cost and revenue is beyond its definition. Revenue planning is perhaps a discrete area that has application to discounted cash flow analysis and overall business budgeting but which is relatively removed from the design of a facility.

Occupancy costs are often the most significant single cost type in a building's life cycle. Despite this fact, occupancy costs are often excluded from the evaluation and analysis processes of design. One reason for this neglect is that these costs are among the most difficult to analyse. Yet this is an area that has the potential to offer significant savings to the client and thus should also be of interest to the design team.

All of the factors comprising occupancy costs are directly or indirectly affected by some components of the design of the building. The most fundamental of these components is simply the floor area. A reduction in floor area would not only contribute marginally to operating cost savings by cutting cleaning and repainting costs as well as energy costs for heating and cooling but also lead to increased efficiency in a facility.

STAFF SALARIES

The largest cost factor influencing the overall cost of a project is staff salaries, yet it is one that is normally ignored by the building economist during design. While the salaries of staff using a facility can be affected by the design team, there are three basic reasons why salaries are not normally considered at the design stage:

- The misconception that design cannot influence expenditure on staff salaries.
- The ability of staff salaries to dwarf the effects of other cost components by comparison. Nevertheless it must be remembered that only a small part of this cost can be eliminated through facility improvement.
- The design team needs to have a high level of knowledge of the functional uses of facilities. Circulation, work patterns, processes and even habits will all have an impact on functions within the building undertaken by staff. Members of the design team generally do not possess an adequate level of this knowledge.

At first thought it may seem that building design and staffing costs have little in common. For example, the shape of an office building may not typically affect the number of employees a firm occupying the building would need to employ.

Staffing here will be dependent on internal company factors such as workload, turnover and strategic planning objectives.

However, consider a building with a more specialised function. A hospital, for example, has very intricate and specialised design requirements. Factors such as supervision and travel distances are important in this context. If these factors can be optimised for each functional area (maybe each ward in this case) then it is reasonable to assume that across an entire hospital the total time savings could result in a reduction in staff numbers (or total work hours). At first it may seem as though the reduction in work hours would be minimal, but when multiplied out each day of the week, each week of the year and each year of the building's life it may accumulate to a significant sum. Staff reductions might also lead to administrative efficiencies and further savings.

However, a hospital is not the only place where design could have an impact on staff quantities. The two key areas of staff activity that design can affect and create savings are:

- Supervision.
- Travel.

These are unproductive activities that are inherent in many operations. Minimising the occurrence of these activities will lead to reductions in operation cycle times and therefore reduced work hours. Even if no staff cuts are made, a saving is still available to the client through the minimisation of unproductive time.

Calculating an estimated value for staff salaries involves estimating the value of two components:

- *Workload* The number and nature of the workers will be required. The estimated number of hours per worker will be the basis for calculating a total cost for staff salaries. If workloads are built-up from the hourly requirements for various activities it is important to make allowances for sickness, leave and the like.
- *Salaries* This is another crucial area that must be approached with caution. The type of worker, the nature of the work and the profitability of the company that will occupy the building will all have an effect on the level of salary. Existing staff profiles should be used as a starting point.

As employee salaries can amass to a significant proportion of a project's total cost, care must be taken in both the calculation and the presentation of any such forecasts to the client. Additionally, a comprehensive sensitivity analysis should be undertaken on the various factors influencing the cost of staff salaries.

OPERATING SUPPLIES

Generally, a component of the functional cost of any project will be expenditure on operating supplies (consumables). The nature of these supplies varies with the function of the building. For example, an office building's operating supplies could include A4 paper, stationery and filing systems. Similarly, bandages in a hospital, sheets in a hotel or food in a prison would all come under the heading of operational supplies.

As a rule of thumb, a design team will not be able to have any significant impact on the level of supply costs, given a fixed operating capacity for a facility. However, there are some exceptions. These can generally be classified as recycling processes. These are often radical and unconsidered aspects of a business, yet as occupancy cost studies span long time horizons, changing future attitudes should be anticipated. Present indications for the future seem to show decreasing availabilities of raw and natural resources, which will send the costs of these materials higher. Also apparent is the increased future dependency on recycled materials as a substitute for these declining raw materials.

Thus it is no longer irrational to consider the inclusion of a recycling facility within a project in order to cut the recurring cost of operational supplies. This proposal, while not commonplace, has actually existed in a few small areas for a long time. For example, glassworks have historically had facilities for purchasing, storing and recycling old glass products. Milk and other beverage co-operatives have had facilities for collecting, washing and reusing bottles for sale in their products.

Increasingly in the future design teams may be called on, by request, necessity or routine, to evaluate the feasibility and compare alternative types of recycling facilities or their components in an attempt to reduce future operational costs.

ADVERTISING

Advertising is the process undertaken by every firm to heighten public awareness of their products and services in order to influence sales. This can take many forms, such as advertisements in the print and electronic media, or sponsorship of non-profit-related activities such as sport, functions or charities. These are all management initiatives which simply involve an outlay of cash and therefore cannot be influenced by any of the physical decisions made by the design team. However, there are two noteworthy exceptions to this principle:

- *Explicit building advertisement* This is simply the incorporation of something additional into the design of the building in order to make the public more aware of the company (e.g. signage on the exterior of buildings). This expenditure can influence the level of sales of an enterprise.
- *Implicit building advertisement* This is based on the premise that the more impressive a building appears, the higher the regard for the company that inhabits it (or has its name on it) by the general public. Improvements in building quality are initially due to the owner's goal of earning increased rent. However, there is also the secondary benefit of increasing public regard by the building's stature.

These two exceptions are factors which the design team can influence and can result in increased sales by the client. Advertising, however, is a process in which it is difficult to predict likely results. It is therefore very hard for an occupancy cost study to quantitatively include costs for advertising or to compare advertising options, since it cannot be stated with confidence that an increase of $X for advertising will result in an annual increase of Y% in sales and a benefit to the client of $Z over the life of the project.

PLANT AND EQUIPMENT

These are costs related to plant and equipment used in the production of the goods and services provided by the building owner. They may include photocopiers in a printing store or lathes in a metalwork factory. As these are generally items that are in most ways isolated from the building as a whole, the design team has virtually no influence on such costs. However, if the design team can obtain knowledge of what plant and equipment may be required then, if considered appropriate, a minimal number of alterations to the design can be made to benefit the client.

Examples could comprise the inclusion of varying voltage dampeners to the electrical installations of a building that will house delicate electrical equipment, analysis of the total wattage output or heat load of machinery in a facility in order to make adjustments to air conditioning, window treatments, etc., adjustments to air conditioning systems if there are humidity-affected stores or materials present, or the correct choice of flooring to minimise wear on mobile equipment.

BUILDING ALTERATIONS

Building repairs and replacement are already included as other forms of operating cost. But building alterations that result from expansion of the function undertaken by a business might need to be anticipated during the design stage.

It is simply common sense that if future expansion or alteration is anticipated and planned for in the initial design, the recurring costs of such alterations will be significantly reduced when it comes time to add another level or erect additional connecting buildings, etc. It is the responsibility of the design team to determine the client's needs and expectations for any project and to make accommodation for any alterations that will need to be made in the future.

WASTE DISPOSAL

As mentioned earlier, recycling will have an increasingly prominent status in the future. However, it is also important as environmental standards are strengthened and become an issue of increasing political and corporate focus. Therefore, another function-related occupancy cost will be that of waste disposal.

Every process produces some type of waste product requiring disposal. At present, deterioration of the environment is a legal and increasingly evident result of overloading natural systems with waste. This is done through exhausts, stormwater drainage, sewer systems and physical litter at dumping sites and tips. As the environment has become burdened and less resilient there have been calls for environmental protection legislation, and a possible outcome of this debate might be a user-pays system in the future.

Perhaps facilities will be obliged by law in the future to treat their own waste. Considerations of this type during the initial design might offer substantial savings later in the building's life.

HEALTHY WORKING ENVIRONMENTS

SBS relates to 'illness associated with the indoor environment where symptoms are non-specific and their causes unknown'. If the cause of the disease is known, such as in the case of Legionnaires' disease, it is called building-related illness (BRI). Together they are having an alarming, and increasing, effect on worker productivity. (Dingle 1995, p. 19)

Absenteeism and reduced productivity due to ill-health are an important consideration when evaluating worker productivity, particularly as the workplace itself can be a primary source of ill-health. Increasing evidence has been gathered over the past two decades to show that subtle problems associated with actual building design and function are having a significant effect on the well-being of inhabitants. This is commonly referred to as the sick building syndrome (SBS).

SBS does not occur in all buildings but is normally associated with buildings used for non-industrial purposes and especially office blocks. Symptoms typically include irritation of the eyes, nose, throat and skin, mental fatigue, reduced memory, lethargy, headaches, dizziness, nausea and unspecific hypersensitivity reactions. Additionally, airborne diseases such as the common cold and flu are easily transferred via conventional airconditioning systems.

The relationship between the individual worker, the design of their workplace and the demands of their job also have an effect on health. Key design features in this area include lighting, workstation design, interaction with other workers and ergonomic design of office furniture and equipment.

The ramifications of a poor work environment do not relate only to reduced productivity levels. Building proprietors and managers are increasingly concerned about the rise in workers' compensation costs and the potential legal consequences of failing to adequately address SBS and other building-related factors that impinge on the health of workers. The future potential costs in this area alone may lead to financial ruin.

HOW TO RESPOND?

This discussion has highlighted some occupancy costs that are within the control of the design team for a project. There are probably many other examples as well. Occupancy costs are significant, but the proportion of those which are subject to optimisation may be relatively small. Nevertheless, this can translate into large sums of money over a reasonable time horizon.

Occupancy cost planning may be an area of increased involvement in the years ahead for building economists. Although detailed knowledge is required, this can be established through research and data collection, which are skills such professionals should possess.

Critics of occupancy cost planning would argue that all the above matters are merely examples of good design. It is the architect's responsibility to examine the functional requirements of the proposed facility, to effectively plan the work environments, minimise circulation and cater for the future needs of the client. But the question remains: what is the cost of these design features and do they represent value for money when examined using a life-cost approach?

The quantification of occupancy costs is likely to remain a rare event for the general run of building work, but in specialised areas such as hospital and hotel design we may find that this activity becomes an area of great potential in the very near future.

REVIEW QUESTIONS

9.1   What is the difference between acquisition and recurrent expenditure?

9.2   What is the result of improvements in supervision and travel/circulation in a building's design?

9.3   What are examples of occupancy costs that can be influenced by building design?

9.4   Why are improvements in worker productivity likely to be so significant?

9.5   What is the difference between explicit and implicit advertising?

9.6   Why might waste disposal and recycling issues be of interest to designers?

9.7   What are denial-of-use costs?

9.8   What is sick building syndrome (SBS)?

9.9   Is society interested in worker productivity?

9.10 Do issues of functional performance overshadow the traditional areas of life-cost study, such as construction, cleaning, energy, repair and replacement expenditure?

TUTORIAL EXERCISE

The refurbishment of an existing supermarket cost $4.5 million. The project includes the introduction of natural day lighting and energy-saving measures designed to reduce the current operating costs of $800 000 per annum by 25%. Following the year in which the supermarket was reopened, sales increased in real terms from $2.1 million per annum to $3.5 million.

What is the payback period for the refurbishment based on operating cost reduction alone? What is the payback period when the extra sales are also included?

> **HINT**
> Payback period is expressed in years.

KEY POINTS

A PowerPoint presentation dealing with the topics discussed in this chapter can be downloaded from the publisher's website (see the publisher's details at the beginning of the book for the address). Some key points are shown below.

Definition

❋ **Occupancy costs are one type of operating cost and hence are a component of total life-cost**

❋ **Occupancy costs can relate to a range of assets, but normally are applied to buildings**

❋ **Costs may include staffing, manufacturing, management, supplies and the like that relate to functional use**

❋ **The term functional-use cost is sometimes used as an alternative descriptor**

Types of Occupancy Costs

❋ Acquisition expenditure comprises outlays for initial purchase of goods and services related to the functional use of the building

❋ Denial-of-use costs, due to delays in acquisition caused by construction overruns, is usually included as part of acquisition expenditure

❋ Recurrent expenditure comprises the ongoing costs of maintaining the functional use of the building

Typical Examples

❋ Examples of acquisition expenditure include office fit-out, computer hardware and software, equipment, permanent advertising signs and other initial capital costs

❋ Examples of recurrent expenditure include staff salaries, manufacturing raw materials, washing and sterilising costs, stationery, advertising, security patrols, replacement equipment and other regular or intermittent operational costs

Significance

❋ A study of a CBD building in Sydney over a 50 year life revealed that 11% of the life-cost was capital cost, 82% was operating cost and 7% was finance costs

❋ Of the 82%, 18% was cleaning, energy, repair, replacement and other miscellaneous expenses and 82% was occupancy costs

❋ Therefore 67% of the total life-cost of the building was related to its functional use

Worker Productivity

❊ Staff salaries are usually the most significant occupancy cost

❊ Unhealthy or uncomfortable work environments can contribute to a decline in the productivity of staff

❊ The major design issues that can affect worker productivity are those associated with visual acuity, thermal comfort, air quality, supervision and travel/circulation

Impact on Design

❊ Day lighting and natural ventilation are design issues that can have a tremendous effect on worker productivity

❊ Floor plan layout and the relationship between functional areas is also important

❊ Sick building syndrome (SBS) results from poor design and can have an alarming effect on building inhabitants, leading to higher staffing needs and absenteeism

Other Considerations

❊ Other examples of occupancy costs that can be affected by design include plant and equipment, operational supplies, advertising, building alterations and waste disposal

❊ Critics argue that efficient levels of occupancy costs are merely the result of good design

❊ But the opportunity to find large savings is present, and even a few per cent gain can represent more than the total annual energy bill

ANSWERS TO REVIEW QUESTIONS

9.1 Acquisition expenditure comprises outlays for initial purchase of goods and services related to functional use, while recurrent expenditure comprises the ongoing costs of maintaining functional use.

9.2 This can reduce the cost of staff salaries by removing unproductive work practices, which can accumulate to a significant proportion of total occupancy costs over a building's life.

9.3 Examples arise from issues concerning floor area, staff salaries, plant and equipment, operational supplies, advertising, building alterations and waste disposal.

9.4 Staff salaries are usually the largest proportion of occupancy costs and even small improvements in productivity can lead to significant savings.

9.5 Explicit advertisement is the incorporation of signage into the building form for promotion, while implicit advertisement is achieved through secondary measures such as total height, use of durable materials and technology to improve image and status.

9.6 As environmental concerns become more prevalent, the costs of waste disposal will improve the feasibility of recycling systems to the point where their integration into buildings is cost-effective.

9.7 Denial-of-use costs are usually included as part of acquisition expenditure and are due to delays in acquisition caused by construction overruns that inhibit the investor from commencing profit-making business activities.

9.8 Sick building syndrome results from poor design and can have an alarming effect on building inhabitants, leading to higher staffing needs and absenteeism through increases in ill-health such as irritation of the eyes, nose, throat and skin, mental fatigue, reduced memory, lethargy, headaches, dizziness, nausea and unspecified hypersensitivity reactions.

9.9 Ultimately society is interest in worker productivity in terms of employment, output, growth and prosperity of the nation.

9.10 Yes, but it should be remembered that it is the proportion of these functional performance costs that may be influenced that is important.

REFERENCES

Dingle, P (1995), 'Sick building syndrome defined', *BEST*, November, pp. 19–21.

CHAPTER 10

SUSTAINABILITY IMPLICATIONS

LEARNING OBJECTIVES
In this chapter you will learn about the true potential of life-cost studies as a valuable technique in the quest for sustainable development. The need to ensure that there is equity between present generations and future generations underpins the use of discounting in comparative life-cost studies. By the end of this chapter you should be able to:
appreciate the importance of environmental values,
- define the principles of sustainable development,
- discuss the ethical characteristics of welfare economics, equity and justice, and
- reflect on possible government policies that can assist in achieving sustainability at the project level.

THE CONCEPT OF SUSTAINABILITY

Sustainable development first appeared in the 1980 World Conservation Strategy.

In the past decade or two sustainable development has been a central concept in the debate about economic progress. Various definitions of sustainable development appear in the literature, since what constitutes development or progress for one individual may not be development or progress for another. Development is a value-related topic and involves judgement on personal ideals and aspirations and the concept of what is in the best interests of society. Nevertheless, a fairly consistent set of characteristics can be identified that define

'The environment should be protected in such a condition and to such a degree that environmental capacities (the ability of the environment to perform its various functions) are maintained over time: at least at levels sufficient to avoid future catastrophe, and at most at levels which give future generations the opportunity to enjoy an equal measure of environmental consumption.' (Jacobs 1991, pp. 79–80)

the conditions under which sustainable development can be achieved.

Development embodies a set of desirable goals or objectives for the progress of society, and these goals undoubtedly include the basic aim to secure a rising level of real income per capita and thereby contribute to an increased standard of living. But there is more to progress than merely income growth. There is now greater emphasis on the quality of life, which includes a variety of factors such as the health of the population, education provision and the care of the natural environment.

The means of achieving sustainable development are:

- *Environmental value* Sustainable development involves a substantially increased concentration on the real value of the natural, built and cultural environments. This higher profile arises either because environmental quality is generally seen as an important factor contributing to the success of the more traditional development objectives such as rising real incomes, or simply because environmental quality is increasingly being viewed as part of the wider development objective and as instrumental in the achievement of an improved quality of life.
- *Futurity* Sustainable development involves a concern not only for the short- to medium-term time horizon but also for the longer term, which will ultimately affect the inheritance of future generations and their quality of life.
- *Equity* Sustainable development places emphasis on providing for the needs of the least advantaged in society (intragenerational equity) and on a fair treatment of future generations (intergenerational equity).

The concepts of environmental value, futurity and equity are integrated into sustainable development through the underlying theme that future generations should be compensated for any reductions in the endowments of resources brought about by the actions of present generations. In other words, if one generation leaves the next with less wealth then it has made the future worse-off. Sustainable development is about making society better-off, and so requires a strategy to ensure that more wealth is left for future prosperity. The definition of wealth, however, requires further investigation.

CAPITAL AND ENVIRONMENTAL WEALTH

Compensation for the future is not achieved by the transfer of money or the creation of debt, but by ensuring that current generations leave succeeding generations with at least as much 'wealth' as the current generation inherited. Wealth can be divided into capital wealth and environmental wealth, representing financial and non-financial qualities respectively. While wealth is ultimately measured in monetary terms, it is purely a convention to enable decisions to be evaluated.

The concept of sustainable development is derivative of the science of environmental economics in several major respects. A fundamental principle is that the economy relies on the natural environment for its future performance. Therefore it is necessary to obtain an understanding of the ways in which the economy and the environment interact. There is an interdependence: the way the economy is managed affects the environment and the resultant environmental quality in turn affects economic performance.

The risk of treating economic management and environmental quality as non-interacting entities has resulted in much of the current environmental crisis. The world could not have continued to use CFCs indiscriminately and cause further destruction of the ozone layer, nor avoided taking action on the reduction of greenhouse gas emissions. Both affect human health and economic prosperity.

'If the nation desires to maintain the capitalized value of its exhaustible resources, it can save a proportion of the annual return from their exploitation so ... after the resources are completely exhausted, the earnings from the savings reinvested in reproducible capital should have the same capitalized value as the initial value of the exhaustible resources and would provide an annual return for all time.' (Mikesell 1977, p. 24)

A common argument in the literature is that development today creates wealth which enriches future generations, enables scientific and technological knowledge to be enlarged and thereby reduces the future's needs for those resources that were depleted in the process. But this is a fallacious view. The concentration on maximisation of capital wealth can result in a reduction in environmental wealth to the extent that future generations are significantly worse-off. Exponential growth is incompatible with sustainable development and so in the longer term must be curtailed. This may involve a reduction in living standards, or at least a reassessment of what constitutes an acceptable standard. As the latter appears preferable to the former, environmental wealth is likely to become more highly valued than capital wealth by future generations.

INTERACTION MODEL

The concept of sustainable development requires that society shift the focus of its environmental policy towards an anticipatory stance. This has implications for the way in which the sustainable development ideal is translated into practice. In the final analysis, however, sustainable development becomes a balance between the concerns of economic progress and environmental conservation. It implies using renewable natural resources wisely so they are not degraded or eliminated and implies using non-renewable natural resources at a rate slow enough to ensure a high probability of an orderly societal transition to new alternatives. The prosperity of future generations in many ways lies within the province of present generations.

Sustainable development, life-cost studies and discounting all affect the environment and contribute to maintain this essential balance. The primary functions of each of these activities demonstrate their interaction with each other and the environment, as shown in Figure 10.1.

Figure 10.1 Interaction Model

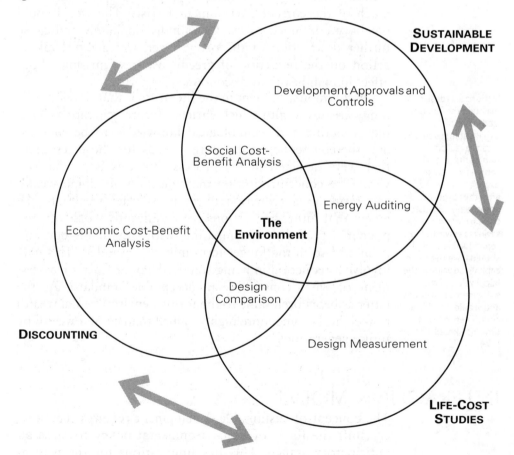

Sustainable development is shown to be primarily concerned with development approval and controls. These are guidelines set down by governments to ensure that both public and private investment occurs within the framework of sustainability. Energy auditing is an area of interaction with life-cost studies and social cost-benefit analysis is an area of interaction with discounting.

Life-cost studies form a suitable technique for the quantification of design performance. This involves the budgeting, optimisation, documentation, monitoring and management of all the costs that flow from a project over its life and the measurement of costs in terms of real value. The comparison of design alternatives is an area of interaction with discounting, and energy auditing is an area of interaction with sustainable development.

Discounting is the methodology that underlies the preparation of an economic cost-benefit analysis in order to select an investment project that delivers the highest return within an appropriate level of risk. Social cost-benefit analysis is an area of interaction with sustainable development, and design comparison is an area of interaction with life-cost studies.

This model forms a useful method of understanding the manner in which discounting, life-cost studies and sustainable development can work together to achieve a common aim.

VALUING THE ENVIRONMENT

One of the central activities in both environmental economics and sustainable development is the placement of proper values on the services provided by the natural environment. Many of these services currently are perceived as having a zero price simply because no marketplace exists in which their true values can be revealed through the commercial actions of buying and selling. Supply and demand theory indicates that if something is provided at a zero price then more of it will be demanded than if it had a positive price. The danger is that a greater level of demand for the environment will be more than natural systems can sustainably support.

It is thus critical that resources and environments which serve economic functions have positive economic value. To treat them as if they had zero value is to seriously risk exploitation and depletion. An 'economic function' is defined as any service that contributes to human well-being, to the standard of living or to development. This simple logic underlies the importance of valuing the environment correctly and integrating those values into a convention (or 'neo-classical') economical framework, and it is in relation to this last objective that discounting assumes great responsibility.

A reactive environmental policy is also referred to as the 'fix it later' syndrome. (Pearce et al. 1989)

Environmental concerns tend to be dissipated by current generations because their impact will occur well into the future and hence rectification can be undertaken at a later time. One reason for postponing action is that future costs are considered less of a burden than current costs. This is reflected in the discounting process, which plays a fundamental role in economic evaluations. Discounting suggests that deferring problems is preferable and this translates to an emphasis on reactive rather than anticipatory policy.

But other considerations work against this purely economic view. Environmental damage can become irreparable (or irreversible) and hence its deference into the future where it will presumably be fixed later might be impossible, even if the relative value of the future dollars used to undertake the repair might be quite small. Furthermore, environmental issues are often treated as if they represent some minor problem in the otherwise efficient working of an economic system that can easily be overcome through theoretical compensation allowances or else ignored completely. Yet the essential feature about environments is that their workings are pervasive and therefore the effects of damage, depletion, pollution, energy usage, waste disposal and the like may manifest themselves in a variety of unexpected ways. Deferring environmental issues to the future therefore attracts a greater level of uncertainty about the nature of the problem, its effect and its ultimate solution.

WELFARE ECONOMICS

Welfare economics concerns the study of the well-being of society as a whole by evaluation of the normative significance of events resulting from the actions of individual members. It must go beyond the bounds of what is called 'positive' economics in that value judgements about utility and amenity are required. Therefore welfare economics differs from the economics of individual consumer behaviour and the behaviour of agencies and firms, being concerned instead with the objectives of society and not the private objectives of individuals.

A substantial part of welfare economics is based on the concept of economic efficiency. This concept relies on two fundamental value judgements to order various social states. The first is that ordering should be based on individual preferences and implies that people are rational and are the best judges of their own actions. This is known as individualism or more commonly consumer sovereignty. The second value judgement is known as the Pareto principle. It states that society is better off if at least some of its members are made better off and no-one is made worse off.

But the Pareto principle is not of great practical use, because it will seldom be satisfied. This gave rise to the Kaldor-Hicks rule, also known as the compensation test, which states that any project should be sanctioned if it improves the welfare of some people, even though others might lose, provided those who gained could compensate those who lost and still have some benefit left over. This rule is the basis of the notion of collective utility (overall social gain) in relation to social cost-benefit analysis.

The objectives of society are not clearly distinct from the aims and ambitions of its members, for if people as individuals are better off then they are better off as a group. But a study of economics based on individual behaviour leaves unanswered some of the most pressing problems concerning social policy. Intergenerational and intragenerational issues are present and are respectively represented by equity and distributional considerations.

Welfare economics is a complex yet important branch of general economics in that it attempts to objectively model what otherwise might be considered as subjective social issues. It derives its tools from basic utility theory and applies them to project decisions so as to assess the benefits that will accrue to society and the policy directions that should be taken.

EQUITY AND JUSTICE

The principle expressed by the sustainability concept itself – that of intergenerational equity – has a number of alternative ethical sources. At its simplest it expounds the Kantian injunction that each generation should do as it would be done by. A more complex interpretation of equity can be derived by asking what distribution of resources would be rational to choose if one were ignorant of the group (or generation) that one belonged to. In both cases there is a refusal to treat other generations as if they are morally less important than the present one and therefore by implication to discount their lives.

Sustainability is anthropocentric: it wishes to preserve the environment for the benefit of future generations. Intergenerational equity is an expression used to describe the balance in the quality of life experienced by society over time – past, present and future. Where it is achieved succeeding generations inherit a quality of life at least equal in standard to that enjoyed by their forebears. In other words, future generations should not be worse off than present generations. Equity in this context does not mean 'equivalence' but 'fairness'.

Gyourko (1991) describes quality of life comparisons undertaken in the United States to differentiate between cities. The relative attractiveness of a city is based on what people are willing to pay in order to live there. Matters considered include climate, culture, employment, home prices, wages, crime and pollution, among others. Generally these types of comparisons are imprecise and are hard to defend.

But surely intergenerational equity is being achieved? People are healthier, they live longer in bigger more sophisticated houses, they have access to better technology, they are better-informed, fly overseas more often and work shorter hours for more pay. On the other hand, crime levels are rising, there are more unemployed and homeless people, pollution is worse and our cities are more crowded.

Listing the factors that can contribute to our quality of life is a long and difficult task. Quantitative measurement is significantly harder. For example, what importance would be placed on such factors as peace on Earth, human rights or global warming? Even if all the factors were identified, ranked and weighted, it is likely that their assessment would differ wildly among individuals.

The sustainable development doctrine requires that present development initiatives not result in a burden for future generations nor diminish their quality of life. When viewed at the project level intergenerational equity becomes much more tangible. The selection process for development must ensure that the balance between initial and recurrent costs or benefits is fair and correctly identifies capital wealth gains. Projects that result in environmental degradation either explicitly by pollution or implicitly by resource consumption need to also value this correctly to identify the effect on environmental wealth.

DISTRIBUTIONAL EFFECTS

Sustainable development implies not simply the creation of wealth and the conservation of resources but their fair distribution. This is primarily an intragenerational issue. Distributional effects involve utility conflict in that some individuals within society may be better off while others are worse off. Although collectively a positive benefit may occur, the disadvantage accruing to 'minority' groups is of concern. In other words, projects should not be based solely on the principle of majority rules.Distributional Effects

In the same manner that equity and justice between generations can be considered by the current valuation of changes in capital and environmental wealth, so too can distributional effects be incorporated in an analysis by the identification of particular social groups and the economic impact that the project will have. The most effective means to overcome distributional effects that are undesirable is financial compensation. Unlike future generations, which are yet unborn, individuals within the present generation can be directly compensated to ensure that fair distribution is achieved.

Life-cost studies would seldom if ever need to address distributional effects, although they are commonly encountered in other forms of economic appraisal, particularly social cost-benefit analysis. In any case, redistribution of benefits is not something that the discount rate can be expected to handle, if for no other reason than that distributional effects are current problems and are not normally related to time adjustments. Yet sustainable development must ensure that poorer communities, such as those in developing countries, are not depleted in order that richer nations may benefit, even in situations where total wealth is enhanced.

REVIEW QUESTIONS

10.1 Is discounting relevant to the evaluation of environmental goods and services?

10.2 How can normal financial analysis take account of environmental impact?

10.3 What is the problem with making investment decisions on the basis of financial return?

10.4 Is sustainable development relevant to developing countries?

10.5 How can sustainable development goals be applied to a specific project in isolation?

10.6   What are the three pillars of sustainable development?

10.7   What is the difference between intergenerational and intragenerational equity?

10.8   What is the Kaldor-Hicks rule?

10.9   What problem does discounting introduce when evaluating projects from a sustainability perspective?

10.10 Does sustainable development exist in reality?

TUTORIAL EXERCISE

Explain how you would go about estimating the value of the damage caused by stormwater pollution to a local waterway. Use some hypothetical figures to indicate an annual value.

Assume the amount of stormwater run-off is 2400 Ml in a typical year.

> **HINT**
> Environmental values are typically estimated using a market surrogate commodity (e.g. air pollution calculated from the number of reported asthma cases multiplied by the cost per patient per annum), an estimate of the cost to rectify the loss (e.g. cleaning up oil-stained beaches), or an estimate of the cost to prevent the loss (e.g. construction concrete barrier walls to absorb traffic noise).

KEY POINTS

A PowerPoint presentation dealing with the topics discussed in this chapter can be downloaded from the publisher's website (see the publisher's details at the beginning of the book for the address). Some key points are shown below.

Definition

* ❋ Sustainable development is the balance between economic progress and environmental conservation, given that both are imperative to our future survival

* ❋ Ecologically sustainable development (ESD) is a term used to mean 'development that meets the needs of the present without compromising the ability of future generations to meet their own needs' (Brundtland Report, 1988)

Concept

* ❋ The environment and the economy necessarily interact

* ❋ Sustainable development involves a substantially increased emphasis on the value of the natural, built and cultural environments

* ❋ Proper values need to be placed on environmental goods and services

* ❋ Achieving sustainable development is a function of environmental value, futurity and equity

Environmental Value

* ❋ Proper values need to be placed on environmental goods and services to avoid overuse and degradation

* ❋ Quality of life needs to be maintained or increased where quality is not merely restricted to monetary wealth

* ❋ Capital and environmental wealth must be considered collectively, and one is not a substitute for the other

Futurity

* Sustainable development involves a concern both with the short-term horizons and the longer-term future to be inherited by our descendants

* This implies not basing decisions solely on current profitability or investment return where future prosperity is being reduced

* Projects should deliver a positive contribution to society when measured over their period of influence after all aspects have been considered

Equity

* Sustainable development places clear emphasis on providing for the needs of the least advantaged in society (intragenerational equity) and on a fair treatment of future generations (intergenerational equity)

* Future generations need to be compensated for reductions in the endowments of resources brought about by the actions of present generations

Fair Compensation

* One view concerning fair compensation is that it is best achieved by ensuring that current generations leave succeeding generations with at least as much capital wealth as the current generation inherited

* But an alternative view suggests that environmental wealth must not be degraded by a focus on maximisation of capital wealth, because living standards are ultimately a function of environmental quality

---

Link to Construction Industry

※ **Development is undeniably associated with construction and the built environment**

※ **Natural resources are consumed by the modification of land, the manufacture of materials and systems, the construction process, energy usage and waste products that result from operation, occupation and renewal**

※ **Buildings are a major contributor to both economic growth and environmental degradation**

ANSWERS TO REVIEW QUESTIONS

10.1   Contemporary literature would suggest no, as the valuation of environmental goods and services is unlikely to decline in a negative exponential (compounding) fashion into the future (if anything, environmental resources will increase in value due to their scarcity).

10.2   For this to occur, environmental issues would need to be embedded into cash flows (not discounted) by taking the total valuation of the gain or loss and dividing it by the time horizon chosen for the analysis (this may be called a 'sustainability constraint').

10.3   Profitability is important, but basing decisions on this criterion alone risks environmental and social loss and is not supported in the sustainable development philosophy.

10.4   Theoretically yes, but an argument can be mounted that developing countries should be given the same opportunity afforded to developed countries in their past to use their natural resources to increase living standards to 'Western' standards.

10.5   The approach is to 'act locally, think globally', but it is acknowledged that global ramifications are far removed by specific project decisions.

10.6   Sustainable development is about proper environmental valuation, futurity (long-term view) and equity (fair treatment).

10.7   Intergenerational equity is fair treatment between present and future (unborn) generations, while intragenerational equity is fair treatment within a generation (e.g. rich and poor).

10.8   The Kaldor-Hicks rule states that a project should be sanctioned if it improves the welfare of some people, even though others might lose, provided those who gained 'could' compensate those who lost and still have some benefit left over.

10.9 The process of discounting diminishes the value of future events, and so encourages a reactive rather than proactive policy stance since the cost of fixing long-term problems is rendered unimportant at high discount rates.

10.10 Not really, it is more of a 'journey' we take than a 'goal' we reach, and as such is similar to other unattainable goals, like perpetual motion and peace on Earth.

REFERENCES

Gyourko, J (1991), 'How accurate are quality-of-life rankings across cities?', *Business Review*, The Reserve Bank of Philadelphia, March/April, pp. 3–15.

Jacobs, M (1991), *The Green Economy: Environment, Sustainable Development and the Policies of the Future*, Pluto Press.

Mikesell, RF (1977), *The Rate of Discount for Evaluating Public Projects*, American Enterprise Institute, Studies in Economic Policy, Washington.

Pearce, DW, Markandya, A & Barbier, EB (1989), *Blueprint for a Green Economy*, Earthscan Publications.

CHAPTER 11

ADDING VALUE

LEARNING OBJECTIVES
In this chapter you will learn about values and the ways in which projects can be improved to deliver greater benefit to society. The role of value management and its relationship to life-cost studies is explored. By the end of this chapter you should be able to:
- understand the importance of focusing on value rather than cost,
- appreciate social responsibilities as a valid project outcome,
- describe typical value management processes applied to construction, and
- measure value for money.

TIME, DISCOUNTING AND VALUE

This section is a critique of Colin Price's 1993 book on time, discounting and value.

In his book *Time, Discounting and Value*, Price concludes that the premises on which discounting is based are mistaken, the logic underlying it misconceived and the case for applying it in practice untenable. Yet the arguments he provides are not totally consistent with this conclusion. Given that the work is an attempt to bring together all of the previous research on discounting, a critique of his findings is both useful and warranted.

Price begins by giving evidence that discounting has been used in the past to calculate equivalent value other than for real cash flows. Among the entities which have been discounted are energy (Hannon, 1982; Ince, 1983; Barrow et al., 1986), employment opportunities (Johnson and Price, 1987), timber volume (Rowan, 1990), levels of radiation exposure (Demin et al., 1983), genetic gain from animal breeding programs (Strandberg and Shook, 1989), carbon-fixing (Price,

1990), electricity output (Shebelev, 1985), machine working hours (Price, 1985), measures of statistical reliability (Young and Ord, 1985) and human lives or years of life saved (Keeler and Cretin, 1983; Cropper and Portney, 1990). The discounting of intangible costs and benefits is not related to investment return and hence is at odds with the use of a capital productivity approach.

Inflation is described as 'a perfectly good reason for discounting the value of money promised some time in the future' (p. 31), and this is undertaken by calculation of real values. Thereafter inflation has little effect on the outcome of decisions and can be effectively ignored. Yet specific inflation can generate a variety of different exchange rates and collectively makes it impossible to determine a single negative exponential discount rate. The general tenor of this viewpoint is accepted.

Internal rate of return (IRR – defined as the discount rate that leads to a net present value of zero) is considered as a possible basis for the determination of the discount rate and is properly rejected. Reinvestment at the IRR is implausible and discounting at such a rate, assuming it can be calculated at all, is conceptually incorrect. The purpose of IRR is to give an indication of overall profitability and to calculate a theoretical break-even point based on maximum return, and clearly has nothing to do with the value of costs and benefits between one point in time and another. IRR is widely advocated (Busby, 1985) and extensively used (Gittman and Forrester, 1977; Dulman, 1989) as a performance criterion, but few suggest that it may also be used as a rate of discount.

Price then considers interest rates as a source of discounting the future. He concludes that 'for individuals who can clearly identify a post-tax rate of return at an acceptable level of risk and liquidity, and who can also include (or are prepared to ignore) environmental and social externalities, that rate of return is appropriate in discounting the value of a specific project' (p. 67). He continues by stating that for companies and government agencies the procedure is similar except perhaps for a different treatment of tax and externalities. This view is accepted and is consistent with the weighted cost of capital approach. Price, however, suggests that the identification of the rate of return is problematic and therefore raises doubts over its use in discounting. This argument appears a difficult one to sustain.

The issue of reinvestment is central to Price's concerns over discounting. He states that whenever the constraints of reality preclude complete reinvestment of early profits discounting loses its relevance, and for those costs and benefits that are intangible (non-monetary) and therefore cannot be reinvested, 'discounting at either internal or general rate of return on investment is indefensible' (p. 83). A different justification is required for discounting non-investible (and non-invested) costs and benefits.

Price's attention then turns to time preference. He states that 'the evidence that people are impatient for early consumption appears overwhelming' (p. 99). However, not all human behaviour conforms with this concept of 'pure' time preference. Impatience is described as inherently irrational, since it is within the sovereignty of the consumer whether to take early consumption or to delay. Human behaviour also is not well represented by a uniform negative exponential, as individuals show strong preference for consumption today compared with next year, but show little discrimination between dates several years ahead (Strotz, 1956). Impatience is more related to 'nowness' than earlier over later, since today is preferred to yesterday in the same way that it is preferred to tomorrow.

Yet despite the fact that few economists deny the existence of impatience (Robinson, 1990), Price rejects it as a reason for discounting. His rationale is due to difficulties of quantification and a disputed incompatibility with negative exponential trends more so than on the basis of conceptual inappropriateness.

The issue of political wisdom is addressed in the context of social time preference. Somers (1971) postulates that it is the political decision-making process that depicts relevant time preference. But politics is arguably more concerned with electoral popularity than long-term endowment and hence raises questions over the validity of political wisdom in respect of the rate of discount. Price argues that even if a government passed the severe tests of knowledgeability, rationality and responsibility there would still be merit in the separate assessment of the discount rate.

The above considerations lead Price to the conclusion that both positive return on investment and impatience 'do not in fact support use of the normal discounting function' (p. 129). Investment return fails to support it because not all returns are reinvested. Impatience fails to support it because behaviour if anything is a function of today over other points in time (past or future) and not one of declining compounded value. He therefore looks elsewhere in search of a valid rationale for discounting. His premise is that weighting the value of future consumption by a uniform negative exponential is an extraordinary process and needs special justification; and if that cannot be found, the process should be stopped.

Specific inflation and escalation are synonymous.

The issue of specific inflation is addressed in the context of physical changes over time. A wide variety of rates of change can be found in the way that goods and services behave, including incremental jumps in price, periods of constant price, declining price and the like. Yet he agrees that many physical and some economic changes can be well described by exponential functions: 'it may be computationally convenient to treat these in the same phase as discounting, adding the relevant figure to or subtracting it from the discount rate' (p. 156). This composite rate of discount is referred to as a quasi-discount rate. The incorporation of specific inflation into the discounting process is, of course, a valid activity: whether or not it is part of the discount rate is a matter of choice, and where adopted in practice it would lead to a number of quasi-discount rates for various commodities.

The discount rate, however it may be determined, can include an allowance for risk. Projects that are more risky would have their future cash flows reduced so that the decision would concentrate more on those costs and benefits that supposedly had the most certainty. Price concludes that, ideally, unpredictable future costs and benefits should be valued by considering the probability distribution of expected outcomes, which can best be handled using a separate risk analysis technique: 'risk should be allowed for ... but discounting is not a good way to do it' (p. 180).

Choice is described as the opposite of risk: under risk an undesired outcome may eventuate against human wishes, under choice it may eventuate in accord with human wishes. Receiving income or gaining control over resources now gives the option to consume or invest, and for this reason is preferable. Again the debate concerns the difficulty of quantification of the benefits of choice and whether it is well represented in the discounting process. While choice and impatience appear to be closely aligned, it is again argued by Price that it is not an effective rationale for the determination of the rate of discount.

Although generally earlier consumption is preferable to later, not all things in life conform to such a simplistic model. For example, expectations and desires may grow over time and become more precious: 'only in exceptional circumstances would response to events and experiences decline exponentially over time' (p. 225). Yet the debate has moved from the tangible to the intangible, and with this shift the discounting philosophy is shown to have less application.

Diminishing marginal utility is identified as a more promising rationale for discounting. This can be viewed in the context of income and the quality of life, or consumption and gross domestic product. A social (market-related) discount rate based on mean growth overemphasises the income growth of the affluent and underemphasises the changing importance of consumption to the poor. This raises issues of intragenerational distribution and equity, as well as equity issues between present and future generations. Price states that 'because of unevenness of distribution, uneven opportunities for income growth, and the implausibility of usual utility of income functions, discount rates based on mean growth rate are systematically biased against those to whom marginal income is most important (the poorest members of society)' (p. 241). Such discount rates obscure the circumstances of individuals in matters such as job promotion, retirement and other similar changes in personal wealth capacity over a normal lifetime.

> Diminishing marginal utility is an economic condition indicating that each additional unit of consumption (or income) is valued less than the previous one. This is an key issue in Price's book.

The issues concerning diminishing marginal utility in relation to income, consumption, environmental resources and sustainability are reviewed at length by Price. The conclusion is drawn that diminishing marginal utility, although conceptually correct and relevant, is nonetheless problematic in its application. Furthermore, the discounting of intangible costs and benefits, such as human life, amenity and environmental considerations, is highly questionable.

Price maintains that nothing intrinsic to the future makes it a worse time for good things to happen, or a better time for bad. Nevertheless, there are sound reasons why the values of certain events would differ over time. But 'the pattern of variation over time may be increasing or decreasing, monotonic or fluctuating, systematic or stochastic, asymptotic to zero, asymptotic to something else, or not asymptotic at all' (p. 289). Only under exceptional circumstances could all this be compiled into a negative exponential function of time.

This conclusion supports the previous notion that the discount rate is made up of project-related, product-related and investor-related components, leading to a number of possible rates (see Chapter 7).

The solution to the dilemma is clearly a more complex approach to the calculation of equivalent value. A single discount rate is not sufficient to account for all the reasons that may cause value to change. This is not a reason to discard discounting. What it indicates is that a number of different rates may be applied to investments: some may be related to the project, some to particular commodities used in the project and others may be related to the investor. Price's real conclusion, therefore, is not that discounting is untenable but that the traditional methods of application oversimplify actual changes in time value. And on this point common ground is achieved.

Price devotes considerable attention to diminishing marginal utility. He agrees that where the realistically expected trend is long-term increase of real income per person, marginal utility of income may be supposed to diminish for individuals and society. But changes in real income based on national averages are inappropriate where the individual or group involved does not demonstrate this rate of growth.

Thus a discount rate based on the mean income growth rate across the community overemphasises income growth of the affluent and underemphasises the changing importance of consumption by the poor: 'where the rich are getting richer and the poor poorer, the social discount rate will be much too high: it may even have the wrong sign' (p. 236). Even where income and its growth are equitably distributed, the derived discount rate is inappropriate for individuals as their income may grow faster than the national average due to their life cycle position, since income growth over a career accelerates and decelerates according to issues such as promotion, years of service, investment, family requirements, retirement and the like.

Price also looks at the impact of diminishing marginal utility on aesthetic resources including the natural environment, buildings and artworks. He suggests that discounting by this rate presupposes that they will become more abundant over time. Technology offers widening opportunities to travel to more distant environmental resources and to use synthetic substitutes for diminishing environmental experiences. However, the peculiar nature of such experiences renders them resistant to substitution. Although buildings and artworks are in increasing supply, they share this peculiar nature. As for those environmental values that are not instrumental to human purposes, diminishing marginal utility is entirely irrelevant to them. Price concludes that the discounting of environmental values is a difficult proposition to sustain.

Furthermore, a utilitarian view of social choice, embodying diminishing marginal utility, does not adequately allow for such indivisible concepts as equity and entitlement. The case for discounting future consumption on grounds of its diminishing marginal utility does not mean that justice dispensed in intertemporal distribution of resources should be discounted. An equity case exists for increasing the allocation of material goods to environmentally impoverished future generations, even if the marginal utility of these is less to them than to the environmentally affluent present. Natural resource limitations coupled with a growing population suggest future shortages and hence increasing rather than diminishing marginal utility. Substitution together with technology advances might still allow increasing consumption of natural resources for many more years. But it is not certain that substitution can proceed to the required degree or that technological advance will occur fast enough or safely enough. The pollutant effects of new technologies may curb their application or require diversion of resources into remedial technology. The success of technology may in fact be the difference between diminishing marginal utility and increasing marginal utility in the future.

Price also considers the satiety of human life and the concept of it being subject to the discounting process as population increases. He concludes that the value to an individual of living as a human being should never be discounted simply because there are more humans on Earth: 'such a viewpoint is ethically repulsive ... and if a narrowly economic theory to discredit it is required, it is simply found ... life is a basket of intramarginal values ... [and] there is no basis for ascribing to it diminishing marginal utility' (p. 287).

Price takes the view that values change at various rates and in opposing directions and cannot be represented by a single negative exponential. Claims that using different discount rates distorts relative values he believes are entirely misconceived. The distortion exists only in relation to the norm of conventional discounting. The rationale for differential discounting is precisely that relative values do change over time.

Much of Price's conclusions on diminishing marginal utility are accepted. There is clear inconsistency between discounting and the value of intangibles, which is sufficient to discredit the process. But Price does not make a convincing case for a similar treatment of tangible costs and benefits.

Like most of the literature, he fails to see that discounting based on interest payments necessary to enable funds to be obtained and used is doing no more than the inclusion of these real costs as a cash outflow. Similarly, the adjustment of changes in real prices (regardless of whether this may be an increase or a decrease) is entirely justified: its incorporation as part of the discounting process is a matter of choice. Real price changes could also be included in the actual cash flow predictions.

Diminishing marginal utility, on the other hand, is introduced so that the balance between present and future prosperity can be modified. But the intention is not to distort or discriminate, but to recognise that changes in prosperity will occur and to enable current decisions to reflect this trend. It is entirely conceivable that diminishing marginal utility may in fact be increasing.

Price completes his exploration of the discounting process by looking at ways in which discounting can be resuscitated. Yet he finds nothing to recommend. He concludes (p. 344):

> It is possible to treat these symptoms by ad hoc constraints and adjustments. But the indwelling malady of uniform discounting will remain. On the other hand, if discounting is abandoned, no further constraint or adjustment is needed, except that environmental disbenefits and opportunity costs must be properly calculated. Fulfilling the requirements of sustainability and recognizing enduring values then simply emerge as norms. Discounting-plus-other-adjustments may sometimes be adopted as an expedient, justified by its results. But ultimately justification depends on the validity of processes. The track of values through time is not generally a negative exponential: we violate the truth whenever we pretend otherwise.

Therefore Price rejects the discounting process. But this is not a satisfactory answer. The proper calculation of values is still as problematic as ever. What is clear is that a single rate of discount is not sufficient to adjust for changing values over time, and hence a more complex approach is required. His conclusions support the idea of discounted future value (see Chapters 6 and 7).

VALUE MANAGEMENT

Value management (VM) is defined as the identification and elimination of unnecessary product cost. Value management is also known by the terms value engineering, value analysis, value improvement, value assurance, value technology and value control. They all have the same general objectives in common: optimising design, function and value. In fact, value engineering is often said to apply to new product design, while value analysis applies to existing products, and both thus become a subset of value management.

At the core of the value management process is the analysis of functions from the point of view of the system as a whole (including the relationship or cost impact of design decisions on the project and/or scheme operation). It is this aspect in particular which distinguishes value management from other methods of improving value.

Function analysis involves clearly and succinctly identifying what things actually do or, perhaps more importantly, what they must do to achieve the project objectives. Through the analysis of functions it is possible to identify wastage, duplication and unnecessary expenditure, thus providing the opportunity for value to be improved. The function analysis perspective enables value management not only to explore the project and/or program brief but also to test the assumptions and needs perceived by the authors of the brief.

Taking a system-wide view has particular relevance to the construction industry. For the most part the construction industry has a compartmentalised approach to the design of facilities. As a result each specialist subgroup is responsible for issuing, reviewing and updating the criteria and requirements of its own specialty. This approach tends to emphasise the performance and costs of the part without due consideration of the performance and costs of the whole.

Value management techniques frequently involve the asking of questions concerning the performance of possible alternatives. These questions include:

- What is it?
- What function does it perform?
- What does it cost?
- What other method or material could perform the same function?
- What would that alternative method or material cost?

The questions and the basic methodology are surprisingly simple, but the technique must maintain sufficient discipline in order to keep the process moving, focused and productive while still supporting spontaneity.

Value management was developed by Lawrence Miles for the US-based General Electric Company in an effort to improve quality and reduce the cost of materials and labour during World War II. He developed an approach through which a product may be improved systematically in both quality and cost-effectiveness. A side benefit was the conservation of limited prioritised resources.

Most of the early value management efforts were aimed at military and other industrial hardware applications. It has since been demonstrated that the basic approach may be successfully applied to construction projects, plant layout, materials handling, organisational analysis and procedures.

Value management became more firmly accepted when the US Navy embraced the technique in the 1950s. They initiated the idea of value management 'teams' of a multi-disciplined background evaluating design and cost in a workshop and seminar setting.

TYPICAL VM APPROACH

Value management is an effective tool for systematically optimising the total cost of a product or activity for a specified period of time in order to arrive at value for money. Value management departs from the standard construction 'cost cutting' approach of simply using cheaper material or reducing quantities, in that it takes into account all expenditures relating to design, construction, maintenance, operation and future replacement and the subjectivity of aesthetics and other matters.

A common theme of modern value management is that a structured, methodical, multi-phased study, conducted over a relatively short period of time, by a multi-disciplined team, generally yields the best results. The typical generic approach used in value management embodies the following phases:

- Information phase.
- Speculative phase.
- Analytical phase.
- Proposal phase.

Two further phases are sometimes included. These comprise the implementation and follow-up phases, and are concerned with ensuring that the proposals generated are put into place and function as intended.

Information Phase

During the information phase, certain questions must be answered. These are:

- What is the item?
- What does it do? (define the function)
- What is the worth of the function?
- What does it cost?
- What is the cost/worth ratio?
- What are the needed requirements?
- What high-cost or poor-value areas are indicated?

Considerable effort, ingenuity and investigation are required to answer these questions. The value management team must determine what criteria and constraints exist at the time of the original design and whether they will continue to apply.

In the information phase the most important steps are (1) determining the basic and secondary functions of the items in the design and (2) relating these functions to cost and worth. The function should be stated in terms that accurately define the problem and at the same time are broad enough to generate the greatest number of alternative solutions. The basic function is the primary purpose of the design. Secondary functions are not required for their own sake – they only augment the basic function. If the design can be changed, the need for the secondary functions may be modified or even eliminated.

To facilitate the evaluation, the function of any item or design can be defined literally by two words: a verb and a noun. Longer definitions are often not concise enough for value management applications. For example, the basic function of a chair is to 'support weight' – support is the verb and weight is the noun. The basic function of a door is to 'provide access'. The function of an electric wire is to 'conduct current'.

A further example illustrates the expansion of thinking that can result. Suppose that there is an area that is infested with mice. Mousetraps have proved inadequate to cope with the problem. The conventional solution might be to 'build a better mousetrap'. However, a more creative way of stating the problem, such as 'eliminate rodents', could lead to solutions that exclude mice from entering as well as methods of killing them.

The next step is to determine the worth of the basic function, which shall act as the 'value standard'. Worth is defined as the lowest cost to perform the basic function in the most elementary manner feasible, within the state of our present technology. Accuracy in estimating worth is not important as it is used only for comparison.

The final step in this phase is to determine what we are actually paying for this function in relation to the value standard. The relationship is called the cost/worth ratio. If it is greater than two or three, poor value and high costs are indicated. The cost/worth ratio gives an indication of the efficiency of the design or item.

Speculative Phase

During the speculative phase the principal question to be answered is: in what alternative ways can the basic function be performed? This phase is designed to introduce new ideas to perform the basic function. Therefore it is necessary to fully understand the problem and, by using problem-solving or creative techniques, to generate a number of ideas that introduce lower-cost alternatives. These additional ideas not only increase the opportunity for cost savings but also enhance optimum solutions for design problems.

The foremost approach to creativity in value management is the brainstorming technique. A brainstorming session is a problem-solving conference wherein each participant's thinking is stimulated by others in the group. The typical brainstorming session consists of four to six people of different disciplines siting around a table and spontaneously producing ideas related to the performance of the required function. During the session the group is encouraged to generate the maximum number of ideas. No idea is criticised. Judicial and negative thinking is not permitted. Evaluation of ideas is deferred until the next value management phase.

The result of a multi-disciplined group in generating ideas has no parallel. The team concept not only results in a large number of ideas but also improves the creative ability of participants. Research tests conducted at the University of Buffalo demonstrated that groups generate 65% to 93% more ideas than individuals working alone.

Analytical Phase

In the analytical phase the value management team examines and then develops the alternatives generated during the preceding phase into lower-cost solutions. The principal tasks are to evaluate, refine and cost-analyse the ideas and to list feasible alternatives in order of descending savings potential.

During this phase the ideas must be refined to meet the necessary environmental and operating conditions of the particular situation. Ideas that obviously do not meet these requirements are dropped. The remaining ideas are potentially workable, subject to the capabilities of present technology, and are cost-analysed. Those showing potential savings are then listed along with their potential advantages and disadvantages. Ideas whose advantages outweigh the disadvantages and which indicate the greatest cost savings are selected for further evaluation. Ways of overcoming the disadvantages need to be discussed.

It is essential that cost evaluation be based on the life-cost of the item rather than its capital cost. This is because value for money is related to a period of usage, not merely initial construction.

After selection of alternatives on a life-cost basis, other constraints, not readily assigned dollar values, must often be considered. Examples of such constraints may include aesthetics, status, noise insulation, environmental impact and the like. A weighted evaluation technique known as 'weighted matrix evaluation' can be employed to assist in the comparative importance of the identified constraints. Those alternatives that satisfy the constraints will receive a higher score than those that do not. Each constraint is later weighted according to its relative importance to the other constraints, and a numeric score for each alternative can be calculated. Commonly constraints are weighted using a scale from one to ten, and each alternative is evaluated using a scale of one to five.

Proposal Phase

The proposal phase is the final step in the value management approach before the recommendations are placed in the hands of 'management'. During this phase three things must be accomplished:

- The team must thoroughly review all alternative solutions being proposed to ensure that the highest value and significant savings are really being offered.
- A sound proposal must be made to management. The team must consider not only to whom it must propose but also how to propose the solutions most effectively.
- The team must present a plan for implementing the proposal. This action is critical: if the proposal cannot convince manage-ment to make the change, all the work amounts to naught.

- The final value management report should include (as applicable):
- A brief description of the project studied.
- A brief summary of the basic and secondary functions.
The results of the function analysis showing existing and proposed designs.
- Technical data supporting the selection of alternatives.
- Life-cost estimates of the existing and proposed designs.
- All associated data, price quotations and suggestions.
- Sketches of 'before' and 'after' designs clearly showing proposed changes.
- A description of the tests or methods used to evaluate the proposed designs.
- A recommendation and ranking in descending order of worth of the proposed changes.

It is important to remember that the value management team can only recommend. It is up to management to concur and the designers to make the final changes.

VALUE FOR MONEY

Therefore value management involves the identification of function and the selection of solutions that can maximise this function at minimum life-cost. Ultimately an effective balance is struck between function and cost and it is this balance that is known as value for money. Commonly life-cost is considered along with other design criteria as a part of the final decision. Each design criterion is weighted as to its importance and the various alternatives being appraised are rated against each criterion using a numeric score. The multiplication of design criterion weight and performance score when totalled for each alternative provides the basis for identification of optimal value. An example of the manner in which subjective and objective issues are collectively analysed is illustrated in Figure 11.1.

Value for money can be determined through division of the value score by the calculated comparative life-cost. The value score represents functional performance issues and is exclusive of matters that can be measured in monetary terms. Life-cost is judged as representing 40% of the decision in the presented example, but this can be altered to reflect various client motives. The higher the value for money index (or benefit ratio), the better is the balance between function and cost. Some form of risk analysis would be undertaken to indicate the probability of the identified value for money being realised.

Figure 11.1 Value Management Evaluation Technique

SUBJECT: External paving for uncovered site paths

PRIMARY FUNCTION: To define and facilitate pedestrian travel over site

DESIGN CRITERIA	weighting (0-10)
A = Safety	10
B = Stability	8
C = Appearance	5
D = Water shedding ability	4
E =	
F =	
G =	

ALTERNATIVES	DESIGN CRITERIA							DESIGN EVALUATION		
scored (0-5)	A 10	B 8	C 5	D 4	E	F	G	VALUE SCORE 60%	LIFE-COST ($/m2) 40%	BENEFIT RATIO
brick pavers on sand base	3 / 30	4 / 32	5 / 25	5 / 20				107	$66.87	2.40
brick pavers on concrete slab	4 / 40	5 / 40	5 / 25	5 / 20				125	$75.55	2.48
concrete pavers on sand base	3 / 30	4 / 32	4 / 20	5 / 20				102	$61.74	2.48
concrete pavers on concrete slaB	4 / 40	5 / 40	4 / 20	5 / 20				120	$70.43	2.56
concrete slab with wood float finish	5 / 50	5 / 40	3 / 15	5 / 20				125	$49.14	3.82
concrete slab with broom finish	5 / 50	5 / 40	4 / 20	5 / 20				130	$49.64	3.93
concrete slab with aggregate finish	5 / 50	5 / 40	5 / 25	5 / 20				135	$67.02	3.02
concrete slab with quarry tile finish	5 / 50	4 / 32	5 / 25	5 / 20				127	$113.03	1.69
concrete slab with cement topping	5 / 50	4 / 32	5 / 25	5 / 20				127	$65.53	2.91
natural stone paving	3 / 30	4 / 32	5 / 25	5 / 20				107	$169.41	0.95
precast concrete blocks	3 / 30	4 / 32	4 / 20	5 / 20				102	$221.40	0.69
pine bark mulching	2 / 20	1 / 8	5 / 25	3 / 12				65	$28.69	3.40
loose aggregate	1 / 10	1 / 8	5 / 25	3 / 12				55	$22.68	3.64
asphaltic concrete	5 / 50	4 / 32	2 / 10	4 / 16				108	$33.44	4.84

Asphaltic concrete is the selected alternative. Loose aggregate is the value standard.

REVIEW QUESTIONS

11.1 Why is VM not just simple cost cutting?

11.2 What is each of the generic VM phases essentially about?

11.3 What is the difference between a primary and a secondary function?

11.4 What is the difference between a secondary function and a performance criterion?

11.5 What is an example of how a properly defined primary function can lead to an innovative solution?

11.6 What is the purpose of a cost/worth ratio?

11.7 Why is the separation of subjective and objective issues of benefit in VM studies?

11.8 How can feedback from VM studies be encouraged?

11.9 Under what conditions would you consider weighting the value score and the life-cost prior to making a recommendation?

11.10 What issues should be considered when attempting to identify areas for targeted VM studies?

TUTORIAL EXERCISE

Using a weighted matrix evaluation, three design choices for the adaptive reuse of an existing inner-city office block have been compared. The life-cost of each alternative has also been calculated over a 25-year time horizon.

Given the following conclusions, which project would you recommend and why? Assume that value score and life-cost are equally weighted

	Value Score (index)	Life-Cost ($/m^2)
Option 1: residential apartments	317	3876
Option 2: retail	250	3088
Option 3: hotel	295	3502

What other considerations need to be incorporated in the analysis?

KEY POINTS

A PowerPoint presentation dealing with the topics discussed in this chapter can be downloaded from the publisher's website (see the publisher's details at the beginning of the book for the address). Some key points are shown below.

Definition

- Value management (VM) is defined as 'the identification and elimination of unnecessary product cost', and is the preferred term
- The technique concerns optimising design, function and value, and is not simply cost cutting
- Value engineering is a term sometimes used to describe new product design, while value analysis is a term sometimes used to describe existing product improvement

Origins of Value Management

- VM was developed by Lawrence Miles for the US-based General Electric Company
- The aim was to improve quality and reduce the cost of materials and labour during World War II
- Most of the early VM efforts were aimed at military and industrial hardware applications
- The US Navy embraced the technique in the 1950s and initiated the idea of multi-disciplined teams and the VM workshop

Generic Phases

* VM in its modern context is a structured, methodical and multi-phased process
* It is broken into generic information, speculative, analytical and proposal phases
* Sometimes two further phases of implementation and follow-up are also included
* Each phase is sequential
* The technique operates best in a multi-discipline environment

Functional Analysis

* Functional analysis is the main activity undertaken in the information phase
* Primary (basic) and secondary functions are identified
* These functions are related to cost and worth
* The worth of the basic function acts as the value standard against which other solutions are compared
* Functions need to be expressed in simple terms

Brainstorming

* Brainstorming is a common activity in the speculative phase
* It consists of a group of people spontaneously producing ideas in an encouraging non-judgmental environment
* Research tests conducted at the University of Buffalo (US) demonstrated that groups generate 65% to 93% more ideas than individuals working alone

Weighted Evaluation

* Weighted evaluation is the main technique applied in the analytical phase

* Alternatives are scored (usually on a scale of 0–5) against a number of weighted performance criteria (usually on a scale of 1–10)

* The multiplication of score and weight is accumulated across all criteria for each alternative

* The highest score indicates the best alternative

Benefits

* A formal VM program can increase profits or savings, provide a positive means for future improvements, improve the image of an organisation, and improve internal operations and communications

* VM is essentially a problem-solving methodology

* Cost reduction does not have to be the basic parameter, as saving time, improving aesthetics, eliminating critical materials, etc. can be pursued

ANSWERS TO REVIEW QUESTIONS

11.1 VM concerns value improvement in the widest sense, and although cost reduction is a key aspect of the technique, it is not achieved by reducing performance or quality.

11.2 The six generic VM phases, in order, concern gathering information, finding alternative solutions, evaluation and ranking, making a clear recommendation, ensuring recommendations can be enacted and providing feedback.

11.3 A primary function is the basic functional activity that the item has to perform, while a secondary function is an additional functional activity that further refines the primary function and must be supported.

11.4 Secondary functions are essential attributes that support the primary function, whereas performance criteria comprise the means of assessing achievement to enable ranking and selection.

11.5 A building is infested by mice, but if 'eliminate rodents' is the primary function rather that 'install mousetraps', solutions aimed at stopping the rodents from entering the building will also be under consideration.

11.6 The cost/worth ratio is the comparison of the cost of a particular solution compared with the lowest cost for provision of the basic function (known as the value standard), so that high ratios (say above 2) indicate potential inefficiency and opportunities for improvement.

11.7 The separation of issues that can be more effectively assessed in monetary terms adds credibility to the process and reflects the balance between functionality and cost, which is at the heart of the value for money decision.

11.8 Post-occupancy evaluation provides a feedback mechanism that can assist future VM studies, and its use can be encouraged by making POE an obligation of the design consultants' conditions of engagement.

11.9 If financial considerations were a driving objective, then it might be appropriate to give greater weight to the life-cost over the value score, and if cost was of little concern, the opposite would apply.

11.10 Probability of significant savings, availability of time and resources, probability of developing alternatives of lower life-cost, and probability of implementation.

REFERENCES

Barrow, P, Hinsley, AP & Price, C (1986), 'The effect of afforestation on hydroelectricity generation: A quantitative assessment', *Land Use Policy*, vol. 3, pp. 141–151.

Busby, RJN (1985), *A Guide to Financial Analysis of Tree Growing*, FAO.

Cropper, ML & Portney, PR (1990), 'Discounting and the evaluation of lifesaving programs', *Journal of Risk and Uncertainty*, vol. 3, pp. 369–379.

Demin, VF, Ermakova, EI & Shebelev, YV (1983), 'Allowance for economic discounting in estimation of the harm done by radioactive contamination of the biosphere by nuclear-energy facilities', *Soviet Atomic Energy*, vol. 54, pp. 207–212.

Dulman, SP (1989), 'The development of discounted cash flow techniques in United States industry', *Business History Review*, vol. 63, pp. 555–587.

Hannon, B (1982), 'Energy discounting', *Technological Forecasting and Social Change*, vol. 21, pp. 281–300.Gittman, LJ and Forrester, JR Jr (1977), 'A survey of capital budgeting techniques used by major US firms', *Financial Management*, vol. 6, no. 3, pp. 66–71.

Ince, PJ (1983), 'COMPARE: A method for analyzing investment alternatives in industrial wood and bark energy systems', *General Technical Report - Forest Products Laboratory*, USDA Forest Service, Washington.

Johnson, JA & Price, C (1987), 'Afforestation, employment and depopulation in Snowdonia National Park', *Journal of Rural Studies*, vol. 3, pp. 195–205.

Keeler, EB & Cretin, S (1983), 'Discounting of life-saving and other non-monetary effects', *Management Science*, vol. 29, pp. 300–306.

Price, C (1993), *Time, Discounting and Value*, Blackwell.

Price, C (1985), 'Capital costing for forest machinery: A discounted output solution', *Scottish Forestry*, vol. 39, pp. 77–84.

Price, C (1990), 'The allowable burn effect: Does carbon fixing offer a new escape from the bogey of compound interest?', *Forestry Chronicle*, vol. 66, pp. 572–578.

Robinson, J (1990), 'Philosophical origins of the social discount rate in cost-benefit analysis', *Milbank Review*, vol. 68, pp. 245–265.

Rowan, AA (1990), *Forest Road Planning*, Forestry Commission Booklet.

Shebelev, IV (1985), 'Using discounted costs to evaluate the effectiveness of economic measures in atomic energy', *Matekon*, vol. 22, pp. 91–110.

Somers, HM (1971), 'On the demise of the social discount rate', *Journal of Finance*, May.

Strandberg, E & Shook, GE (1989), 'Genetic and economic responses to breeding programs that consider mastitis', *Journal of Dairy Science*, vol. 72, pp. 2136–2142.

Strotz, RH (1956), 'Myopia and inconsistency in dynamic utility maximization', *Review of Economic Studies*, vol. 23, pp. 165–180.

Young, P & Ord, JK (1985), 'The use of discounted least-squares in technological forecasting', *Technological Forecasting and Social Change*, vol. 28, pp. 263–274.

CHAPTER 12

HOLISTIC PROJECT APPRAISAL

LEARNING OBJECTIVES
In this chapter you will learn about multi-criteria analysis as a technique for taking a holistic view of project selection. The key criteria are investment return, functional performance, energy usage and loss of habitat. By the end of this chapter you should be able to:
- assess the four key criteria using appropriate units of measure,
- balance these criteria to choose those projects that represent the highest value,
- benchmark projects and compare performance, and
- calculate a sustainability index.

SUSTAINABLE PROJECT EVALUATION

In order to make a real difference in the future, the development process must become more sympathetic to sustainability ideals. It is unlikely that this will occur of its own accord across the full range of built infrastructure. If environmental quality is to be protected, more rigorous development controls need to be applied and proper analytical tools must be used that are capable of integrating project feasibility and project sustainability criteria. This must be supported by the progressive introduction of minimum performance standards for completed facilities, a system of self-assessed compliance, and effective enforcement and rectification procedures. The current conflict of interest between investor-centred and community-centred objectives must be resolved through the introduction of enhanced approval processes.

MULTI-CRITERIA ANALYSIS

In normal life people are faced with decisions on a regular basis. The process of decision-making involves identifying, comparing and ranking options using multiple criteria. Often this process occurs without even thinking much about it. However, for large investment-related problems there is a tendency to simplify the decision process into a single monetary criterion. The selected option is the one that demonstrates the highest monetary value.

Project appraisal techniques are employed to structure the diverse array of data into a manageable form and provide an objective and consistent basis for choosing the best solution for a situation. Cost-benefit analysis (CBA) is a respected appraisal technique that is widely used by both public and private organisations to aid in complex investment decisions. In conventional CBA much effort has been made to assess the input costs and output benefits by means of a market approach. With the increasing awareness of possible negative external effects and the importance of distributional issues in economic development, the usefulness of CBA has become increasingly controversial. In response to this dissatisfaction, recent attention has been paid to multi-dimensional evaluation approaches.

The identification of value for money on development projects is still related to monetary return. But other issues are also relevant, particularly for social infrastructure projects, and some are becoming increasingly significant. For example, issues such as functionality and resource efficiency are vital to the assessment of sustainable development in the wider social context. Since no single criterion can adequately address all the issues involved in complex decisions of this type, a multi-criteria approach to decision-making offers considerable advantage.

Social costs and benefits (including those related to environment impact) need to be integrated into the evaluation process and a strategy developed that will give these factors proper consideration in practice. Social costs and benefits should not be discounted alongside conventional cash flows, as they bear little relationship to financial matters and do not

Project feasibility is often confused with project sustainability. Actually the two concepts have little in common. A project can be feasible because it generates a high benefit-cost ratio (at least greater than 1), but may be quite unsustainable in terms of its impact on the environment, its use of energy, and its contribution to community expectations including a whole raft of non-monetary considerations. On the other hand, a project can have minimal impact on its surroundings, be low-maintenance, use solar power for heating and cooling, recycle all waste and generally be self-sufficient, yet be totally unfeasible as a financial investment.

reduce in importance exponentially over time. In fact, future generations may value environmental issues more highly than the present generation.

A multi-criteria approach is better suited to deal with the complexities of sustainability assessment. Key criteria such as investment return, functional performance, energy usage and loss of habitat can be combined to form a single decision model that can rank options according to the perceived level of achievement of sustainable development objectives. Such an approach moves away from conventional economic evaluation and the inherent problems associated with monetary assessment.

Traditional CBA uses price as the main tool to evaluate projects based on market transaction. However, over the past decade criticism has stemmed from attempts at putting underlying welfare economic theory into practice. It is often difficult or even impossible to improve social welfare in a society if the natural environment continues to be consumed and depleted. The CBA framework ignores and/or underestimates environmental assets as there are considerable difficulties in measuring all relevant impacts of a project in monetary units.

Intangible and externalities have become major components of concern in project development decisions, in particular due to the possibility of undesirable effects on the natural environment. The presence of externalities, risks and spillovers generated by project development preclude a meaningful and adequate use of a market approach methodology. When the analysis turns to such effects as environmental quality and loss of biodiversity, it is rarely possible to find monetary variables that can provide a valid indicator. Although effort has been made to arrive at values for social costs and benefits, in practice it is almost impossible to place anything more sophisticated than arbitrary numerical values on such effects. The requirement for incorporating environmental issues into project appraisal process is increasing, and so the imputation of market prices becomes more and more questionable.

Equity is another consideration in which market evaluation methods perform poorly. Equity distribution is largely overlooked because it is just the total sum of monetarised effects that is used to determine acceptance. The outcome of development can benefit some groups in the community while others are confronted with negative effects. The methodology assumes that so long as a project generates positive return, the groups disadvantaged by the decision can be compensated. However, no real compensation necessarily takes place in practice. Thus from a social welfare and justice point of view the equity issue has to be dealt with in another way.

For more information on the rationale for MCDM, see Nijkamp et al. (1990), van Pelt (1993), Hanley (1992) and Abelson (1996).

Alternatives have been developed to replace CBA completely with other techniques that do not require social costs and benefits to be monetarised. Cost-effectiveness analysis (CEA) and environmental impact assessment (EIA) are leading solutions in this respect. Others have suggested supplementing CBA with a technique that can measure environmental costs in different ways. Multiple criteria decision-making (MCDM) is widely accepted to aid evaluation processes.

MCDM is a technique designed to value two or more criteria and it is particularly useful for those social impacts that cannot be easily quantified in normal market transactions. MCDM transfers the focus of measuring criteria with prices to applying weights and scores to these impacts to determine a preferred outcome. The total scores that reflect the importance of environmental impacts are used to rank project options. MCDM is a more flexible methodology and can deal with quantitative, qualitative and mixed data, while CBA may handle only data of a quantitative nature. In contrast to CBA, MCDM does not impose any limitation in the number and nature of criteria. Therefore, MCDM is a more realistic methodology in dealing with the growing complexity of development decisions.

SINDEX: A SUSTAINABILITY MODELLING TOOL

SINDEX is a Microsoft Visual Basic.net program developed originally by Computerelation Australia Pty Limited and available via the author of this book as a teaching tool. It is sold under an educational site licence.

SINDEX is an easy-to-use computer-based tool to calculate and benchmark sustainability performance using a multi-criteria approach. The framework was developed to assist practitioners who are charged with making judgements and giving advice concerning the contribution made by new buildings, existing facilities and indeed any product or asset. Sustainability is defined as a balance between predominantly economic criteria (investment return and energy usage) and social criteria (functional performance and loss of habitat). The framework on which this software is built aims to maximise wealth and utility and minimise resources and impact, and thereby to create an opportunity to trade off one with another. Importantly, financial prosperity is part of the sustainability calculation and contributes to a balanced solution. Benchmarks are set across projects to enable accept/reject decisions to be made as well as relative ranking. Overall, the derived sustainability index should be greater than or equal to 1.

When evaluating projects and facilities it is important to take a holistic view. Current practice, however, is primarily concerned with issues of profitability and the financial bottom line. This approach leads to decisions that are not necessarily compatible with wider social considerations and sustainable development goals. A new methodology is needed.

John Elkington proposed the triple bottom line concept in 1997. This approach demands consideration of financial, social and environmental parameters (known as the 3Ps of profit, people and places). It is an approach that is receiving widespread international recognition. Some people advocate a fourth parameter (ethics) to deal with issues of intergenerational equity. Such methodologies are examples of multi-criteria decision analysis. Using recognised techniques, various criteria can be measured and assembled into a single decision model. In this way it is possible to compare alternatives to determine best value, and to benchmark projects and facilities against established performance targets.

SINDEX uses multiple criteria to calculate a sustainability index, which potentially replaces conventional net present value methodologies for ranking and selecting projects. The work has clear international application. Based on a literature review and survey, key objectives were narrowed down and grouped into the four criteria and summarised as:

- maximise wealth (investment return)
- maximise utility (functional performance)
- minimise resources (energy usage)
- minimise impact (loss of habitat)

Profitability is considered part of the sustainability equation. The objective is to maximise wealth. Investment return is measured as benefit-cost ratio (BCR) and includes all aspects of maintenance and durability.

Functional performance, including social benefit, is another clear imperative. Designers, constructors and users all want to maximise utility, which can relate to wider community goals. A weighted evaluation matrix can be used to measure utility in a quantitative manner.

Resources include all inputs over the full life cycle, and can be expressed in terms of energy (embodied and operational). Resource usage needs to be minimised as much as possible. Energy can be measured as annualised GJ or GJ/m^2.

Loss of habitat encompasses all environmental and heritage issues. The aim is to minimise impact. Assessment scorecards (questionnaires) are a useful method to quantify impact, and can be expressed as a risk probability factor.

MAXIMISE WEALTH (investment return) is an economic criterion. The software uses a discounted cash flow approach to determine the net present value and benefit:cost ratio based on tangible costs and benefits over a 30-year time horizon (see Figure 12.1). The choice of time horizon is fixed so as not to encourage short-term decision-making or unrealistic long-term estimates. A user-defined discount rate adjusts the cash flow stream. The default discount rate is 3%, but typically valid rates may range between 0% and 6%, although a wider range is selectable. The residual (scrap) value at the end of 30 years should not be overlooked.

Costs and benefits should be entered as whole numbers without decimal places, and particularly without commas or other non-numeric characters. Costs include acquisition (Year 0) expenses and cleaning, operating, maintenance and replacement costs each year. Benefits comprise income or other inflow, and may include tax concessions if desired. All values should be in present-day (real) terms.

Investment return

Figure 12.1 Maximise Wealth Input Screen

	Cost	Benefit	Net Benefit	Discounted Net Benefit
Maximize Wealth (Discounted Cash Flow)				
Year 0	50000000	0	-50000000	-50000000
Year 1	100000	2000000	1900000	1862745
Year 2	100000	4500000	4400000	4229143
Year 3	100000	4500000	4400000	4146219
Year 4	100000	4500000	4400000	4064919
Year 5	250000	4500000	4250000	3849356
Year 6	100000	4500000	4400000	3907074
Year 7	100000	4500000	4400000	3830465
Year 8	100000	4500000	4400000	3755358
Year 9	100000	4500000	4400000	3681723
Year 10	750000	4500000	3750000	3076306
Year 11	100000	4000000	3900000	3136626
Year 12	100000	4000000	3900000	3075124
Year 13	100000	4000000	3900000	3014827
Year 14	100000	4000000	3900000	2955712
Year 15	250000	4000000	3750000	2786305
Year 16	100000	4000000	3900000	2840938
Year 17	100000	4000000	3900000	2785234
Year 18	100000	4000000	3900000	2730621
Year 19	100000	4000000	3900000	2677080
Year 20	1500000	4000000	2500000	1682428
Year 21	100000	4500000	4400000	2903013
Year 22	100000	4500000	4400000	2846092
Year 23	100000	4500000	4400000	2790286
Year 24	100000	4500000	4400000	2735575
Year 25	250000	4500000	4250000	2590506
Year 26	100000	4500000	4400000	2629349
Year 27	100000	4500000	4400000	2577793
Year 28	100000	4500000	4400000	2527248
Year 29	100000	4500000	4400000	2477694
Year 30	750000	4500000	3750000	2070266
Residual		40000000	40000000	22082836

Discount Rate = 2   Net Present Value = 62318861

BCR = 2.145302

Functional performance

MAXIMISE UTILITY (functional performance) is a social criterion. The methodology employed is known as a weighted evaluation matrix and is commonly used in value management studies (see Figure 12.2). Utility is defined as functional performance and is intended to include non-monetary (intangible or external) benefits. Examples of performance criteria that may be employed are aesthetics, flexibility, community benefit, adaptive reuse potential, noise privacy and indoor air quality. Financial considerations such as durability and ease of construction should not be used. Performance criteria are weighted on a scale of 1 to 10 (10 being the most important) to reflect relative priorities. Each performance criterion is scored using a scale from 0 to 5 (0 being unacceptable and 5 being excellent) to indicate satisfaction levels. The higher the total value score, the more attractive the project is perceived to be. Up to 12 performance criteria can be used. Evaluation is based on the calculated score compared to the maximum score possible.

Figure 12.2 Maximise Utility Input Screen

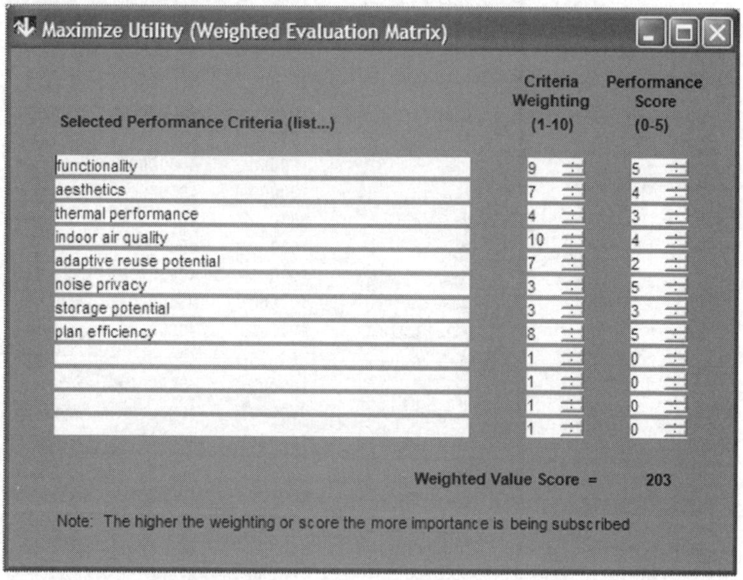

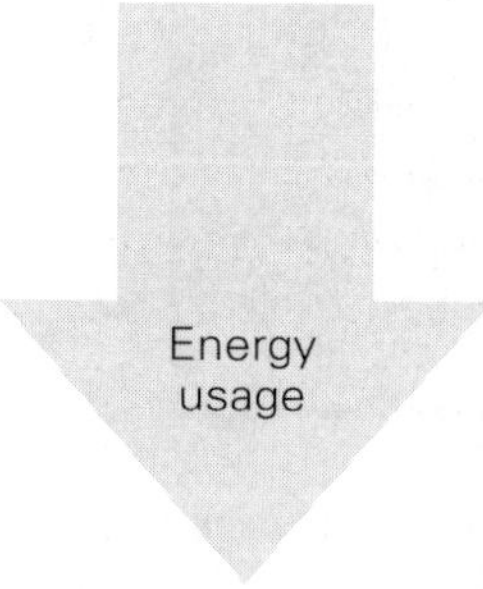

MINIMISE RESOURCES (energy usage) is also an economic criterion. Resources are defined as a reflection of both embodied and operating energy expressed on a per annum basis (see Figure 12.3). Embodied energy includes upstream processes in material manufacture such as mining, processing and delivery. Operating energy includes fuel non-renewable alternatives. Actual (or expected) energy is compared to a target based on either normal practice or legislated limits, and where actual exceeds target the percentage that needs to come from renewable resources is calculated.

All energy values are entered as GJ or GJ/m^2 in whole numbers without decimal places, commas or other non-integer characters. It should be noted that while the cost of materials and energy is included in the previous criterion, this is not double-counting the impact of energy but rather ensuring that efficient use of resources remains an important goal.

Figure 12.3 Minimise Resources Input Screen

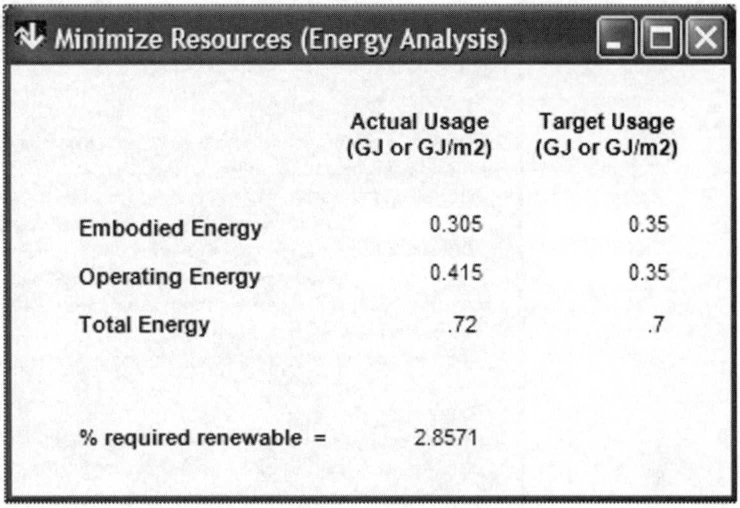

⇓ Minimize Resources (Energy Analysis)	⬜⬜✖	
	Actual Usage (GJ or GJ/m2)	**Target Usage (GJ or GJ/m2)**
Embodied Energy	0.305	0.35
Operating Energy	0.415	0.35
Total Energy	.72	.7
% required renewable =	2.8571	

Loss of habitat

MINIMISE IMPACT (loss of habitat) is also a social criterion. A risk assessment questionnaire is used to gauge the likelihood of damage to environmental assets in terms of loss of biodiversity, global warming, ozone depletion, pollution and other detrimental (often global) effects, as well as heritage considerations. The questionnaire is divided into manufacture, design, construction, usage and demolition sections, with an optional context section where the project involves a physical site (see Figure 12.4). The responses expected are largely qualitative but can be reasonably derived from a general understanding of the project across its entire life cycle.

The calculation performed is intended to determine a risk level ranging from 0 to 20% (minimal impact) through to 80 to 100% (unacceptable impact). Obviously, the lower the risk the better in this case. While risk does not necessarily translate into reality, it gives a reasonably good indication of potential environmental damage, which may be offset by reviewing decisions incorporated in other criteria.

Figure 12.4 Minimise Impact Input Screen

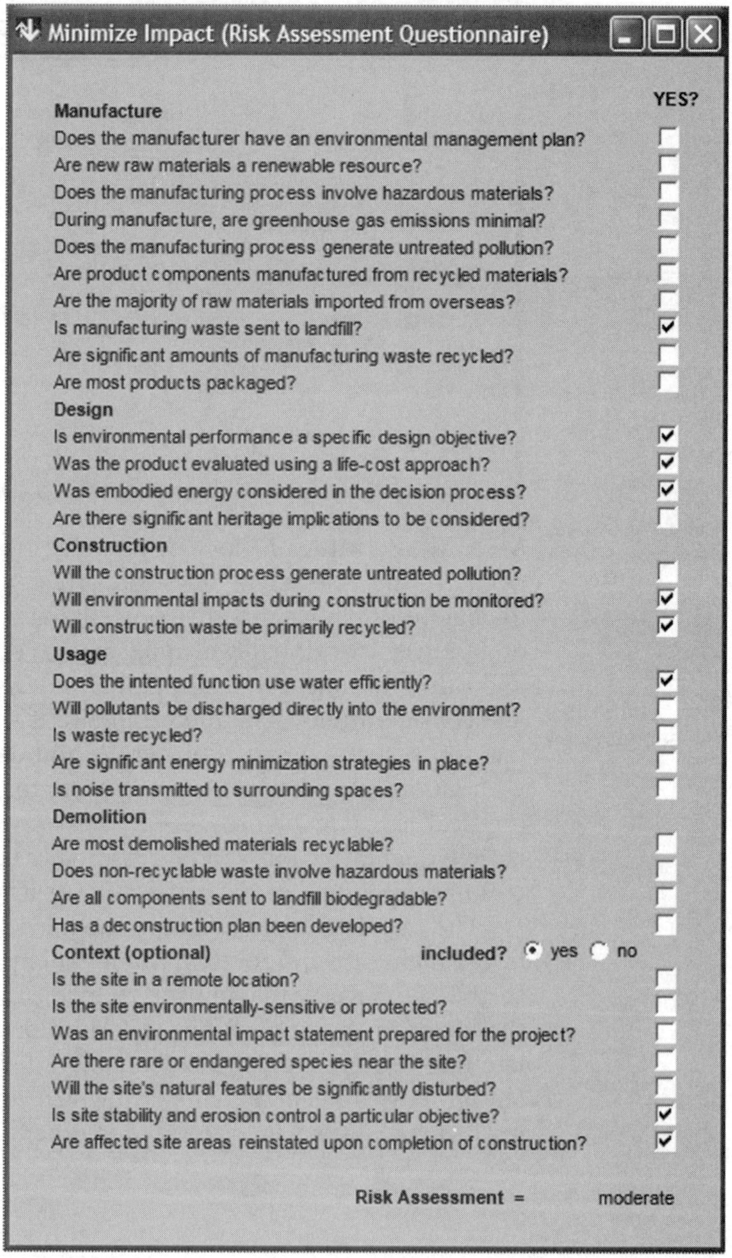

Minimize Impact (Risk Assessment Questionnaire)

	YES?
Manufacture	
Does the manufacturer have an environmental management plan?	☐
Are new raw materials a renewable resource?	☐
Does the manufacturing process involve hazardous materials?	☐
During manufacture, are greenhouse gas emissions minimal?	☐
Does the manufacturing process generate untreated pollution?	☐
Are product components manufactured from recycled materials?	☐
Are the majority of raw materials imported from overseas?	☐
Is manufacturing waste sent to landfill?	☑
Are significant amounts of manufacturing waste recycled?	☐
Are most products packaged?	☐
Design	
Is environmental performance a specific design objective?	☑
Was the product evaluated using a life-cost approach?	☑
Was embodied energy considered in the decision process?	☑
Are there significant heritage implications to be considered?	☐
Construction	
Will the construction process generate untreated pollution?	☐
Will environmental impacts during construction be monitored?	☑
Will construction waste be primarily recycled?	☑
Usage	
Does the intended function use water efficiently?	☑
Will pollutants be discharged directly into the environment?	☐
Is waste recycled?	☐
Are significant energy minimization strategies in place?	☐
Is noise transmitted to surrounding spaces?	☐
Demolition	
Are most demolished materials recyclable?	☐
Does non-recyclable waste involve hazardous materials?	☐
Are all components sent to landfill biodegradable?	☐
Has a deconstruction plan been developed?	☐
Context (optional) included? ⦿ yes ◯ no	
Is the site in a remote location?	☐
Is the site environmentally-sensitive or protected?	☐
Was an environmental impact statement prepared for the project?	☐
Are there rare or endangered species near the site?	☐
Will the site's natural features be significantly disturbed?	☐
Is site stability and erosion control a particular objective?	☑
Are affected site areas reinstated upon completion of construction?	☑

Risk Assessment =   moderate

These four criteria can be assembled together to illustrate the performance of new projects and changes to existing facilities. This approach creates a rating tool to predict the extent to which sustainability ideals are realised, but it also acts as an aid in ongoing facility management processes. Criteria can be individually weighted to reflect particular client motives. Value for money is defined as the ratio of investment return to resource input. It is investor-centred. The higher the ratio, the more attractive the option becomes. Standard of living is more community-centred. It can be measured as the ratio of utility to impact, and includes externalities. High ratios are again preferred.

When all four criteria are combined, an indexing algorithm is created that can rank projects and facilities on their contribution to sustainable development. The algorithm is termed the *sustainability index*. Each criterion is measured in different units reflecting an appropriately matched methodology. Criteria can be weighted to give preference to investor-centred or community-centred attitudes and benchmarked against best practice.

To complete an evaluation, it is normal to click on the four 'pieces' of the sustainability puzzle and complete the required fields. As this is done, a sustainability index is displayed on the main page (see Figure 12.5). The calculation algorithm can be affected by a user-defined emphasis between economic criteria (akin to measuring value for money) and social criteria (akin to measuring living standards). It is recommended, however, that the range of adjustment be limited to 25:75 or 75:25 so that all four criteria play a part in the final outcome. The default is 50:50, which means all criteria are equally weighted. A characterised face is employed to give an overall summary of the outcome (see Figure 12.6), and moves through five expressions from 'crying' (an unsustainable project) to 'excited' (a well-balanced project). While the sustainability index can be any positive value, it is usually in the range of 0 to 5 – where the higher the index, the better the result.

Benchmarks are preset for each criterion. Calculated values for each criterion can therefore be compared to these benchmarks. While the general rule is that a sustainability index of 1 is considered as the basis for accept/reject decisions, some projects may be ruled out if any of the four criteria fall below set benchmarks. More probably, however, where this occurs, ways should be investigated to remedy the situation while still keeping the index as high as possible. This is the challenge for both designers and owners. The preset benchmarks are BCR=1 for maximise wealth, 'actual' equals 'target' for minimise resources, 50% of the maximum score for maximise utility and 50% risk exposure for minimise impact. A comparison to these benchmarks is illustrated on the main page by way of a 'tick' or 'cross'. Benchmarks serve to ensure that similar projects can be compared and evaluated and regulatory authorities can specify minimum performance thresholds they determine are in the national interest.

Figure 12.5 Main Page

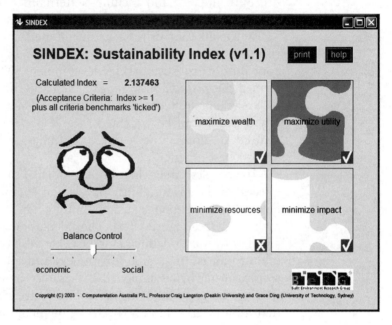

Figure 12.6 Summary Icons

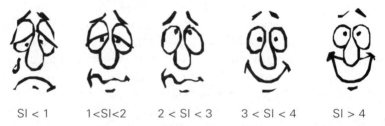

| SI < 1 | 1<SI<2 | 2 < SI < 3 | 3 < SI < 4 | SI > 4 |

In order to make a real difference in the future, the development process must become more sympathetic to sustainability ideals. It is unlikely that this will occur of its own accord across the full range of built infrastructure. If environmental quality is to be protected, more rigorous development controls need to be applied and proper analytical tools must be used that are capable of integrating project feasibility and project sustainability criteria. This must be supported by the progressive introduction of minimum performance standards for completed facilities, a system of self-assessed compliance, and effective enforcement and rectification procedures. The current conflict of interest between investor-centred and community-centred objectives must be resolved through the introduction of enhanced approval processes.

This is a challenge for the twenty-first century.

REVIEW QUESTIONS

12.1 What is meant by triple bottom line?

12.2 What is the essential feature of multi-criteria analysis?

12.3 What are the four criteria used in the sustainability index and what units are they measured in?

12.4 Is there overlap between performance criteria?

12.5 What is meant by investor-centred and community-centred attitudes?

12.6 What does a sustainability index of 0.9 mean?

12.7 Why are weightings necessary within the index algorithm and what restrictions should be observed?

12.8 Can the sustainability index completely replace conventional selection criteria?

12.9 What happens where projects do not meet criterion benchmarks?

12.10 Should taxation be considered in calculation of the BCR?

TUTORIAL EXERCISE

You have been asked to evaluate six proposed options for investment using the multiple criteria of wealth, utility, resources and impact.

Option	Wealth (BCR)	Utility (value score)	Resources (actual GJ/m^2)	Impact (risk score)
A	1.41	350	0.50	25
B	3.29	210	0.60	19
C	1.15	275	0.60	18
D	0.87	298	0.40	17
E	2.61	290	0.55	23
F	4.20	130	0.80	4

After an initial analysis you have estimated values for each criterion. Given that all four criteria are to be equally weighted, what is your recommendation about how to proceed? Test your recommendation for a bias towards value for money (delivered wealth per unit of resource) and quality of life (delivered utility per unit of impact). In both cases assume that the bias is in the ratio of 75:25.

Following a similar methodology to that previously described for SINDEX, benchmarks for suitable compliance have been set as follows:

- wealth: benefit-cost ratio > 1
- utility: value score > 50% of maximum score
- resources: actual GJ/m^2 < target usage
- impact: risk probability < 50%

The maximum value score in this comparison is 400. The energy target usage is set at 0.65 GJ/m^2. Risk score is out of assessed out of 33, where a score of zero represents maximum impact. If any project does not meet the above benchmarks, it should be eliminated from the comparison. In reality, such problems might lead to significant redesign.

> HINT
> Use the following formula to calculate the sustainability index:
>
> SI = value for money + quality of life
>
> $$= \frac{\text{investment return}}{\text{energy usage}} + \frac{\text{functional performance}}{\text{loss of habitat}}$$

KEY POINTS

A PowerPoint presentation dealing with the topics discussed in this chapter can be downloaded from the publisher's website (see the publisher's details at the beginning of the book for the address). Some key points are shown below.

Multi–Criteria Analysis

❋ Project appraisal dictates multiple performance criteria

❋ Criteria are measured using different units that are appropriate for assessment

❋ This approach is not dissimilar to John Elkington's triple bottom line

❋ Criteria can be individually weighted to reflect particular client motives and combined into a single index

---

Four Criteria

❋ To evaluate built projects and facilities, four criteria have been selected

❋ These were derived from an extensive industry survey

❋ The criteria comprise investment return, functional performance (including community benefits), energy usage and loss of habitat

❋ Two criteria need to be maximised and two criteria need to be minimised

Investment Return

❋ Investment return includes all tangible costs and benefits that flow from the project

❋ Costs include land, construction, cleaning, energy, maintenance and replacement expenditure

❋ Benefits include income through rent, lease, sale and advertising

❋ Taxation implications should be part of the calculation

Functional Performance

❋ Functional performance includes all intangible benefits that do not result in a direct transfer of money

❋ Examples include functionality, aesthetics, flexibility, efficiency, safety and adaptive reuse potential

❋ Social and community objectives should be considered here

❋ Durability is best treated as a tangible cost

Energy Usage

* Energy usage comprises all resources consumed by the project, both initially (embodied) and subsequently (operating)

* Examples include electricity, natural gas, petroleum, coal and all renewable fuels

* Energy usage is expressed in GJ or GJ/m2 and is not a measure of energy cost

* Embodied energy includes all upstream processes in mining, manufacture and delivery

Loss of Habitat

* Environmental concerns are included under this criterion

* Loss of habitat refers to the disruption that the project causes to existing eco-systems

* By definition this comprises both local and global impacts, although the latter are clearly linked to individual projects

* Heritage and cultural losses are to be included here

Sustainability Index

* The sustainability index is defined as the balance of all four criteria, expressed as a single value

* Criteria can be weighted or considered equal

* This index can replace traditional selection criteria such as net present value and benefit-cost ratio

* A sustainability index greater than 1 indicates that the project is worthwhile, and the higher the index the better

ANSWERS TO REVIEW QUESTIONS

12.1 The triple bottom line is an accounting convention developed by John Elkington in 1997 to encourage business to consider social (people) and environmental (places) in addition to financial (profit) issues.

12.2 MCA enables multiple criteria to be measured in different units appropriate to each criterion and assembled into a single value for interpretation.

12.3 The four criteria are investment return (benefit-cost ratio), functional performance (weighted value score), energy usage (GJ/m^2) and loss of habitat (risk probability).

12.4 No, the two criteria to be maximised deal with tangible and intangible benefits respectively, while the criteria to be minimised deal with resource and habitat usage.

12.5 Investor-centred attitudes refer to profit motives (value for money), while community-centred attributes refer to social contribution (quality of life).

12.6 This means the project, on balance, has not met the acceptability threshold (SI=1) and therefore should either be dropped or redesigned.

12.7 Some projects have an economic bias and others have a social bias, so weightings are used to recognise this diversity, yet it is important that all criteria are part of the index in all cases (i.e. minimum 25% weighting).

12.8 Yes, the sustainability index is used to rank and judge projects on their contribution to society and prevents projects from being promoted on the basis of narrow goals and responsibilities.

12.9 Regardless of the value of the index, any criterion that does not meet the stated benchmarks leads to the project being dropped or redesigned.

12.10 Yes, taxation issues need to be incorporated in the cost and benefit cash flows (expressed in future value terms) and discounted by a rate comprising after-tax investment return less inflation.

REFERENCES

Abelson, PW (1996), *Project Appraisal and Valuation of the Environment: General Principles and Six Case-Studies in Developing Countries*, Macmillan.

Hanley, N (1992), 'Are there environmental limits to cost benefit analysis?' *Environmental and Resource Economics*, vol. 2, pp. 33–59.

Nijkamp, P, Rietveld, P & Voogd, H (1990), *Multicriteria Evaluation in Physical Planning*, Elsevier North-Holland.

van Pelt, MJF (1993), *Ecological Sustainability and Project Appraisal*, Averbury.

APPENDIX 1

LIFE-COST CALCULATIONS

PURPOSE
This section contains an example of both the comparison and measurement of life-costs. It is presented here to highlight the vast difference between these two activities, particularly in regard to the relevance of discounting. All data are hypothetical and for illustrative purposes only.

COMPARISON OF ALTERNATIVES

When comparing alternatives, discounting is used to account for changes in the time value of money. In addition to matters of opportunity, the discount rate includes a factor for expected affordability fluctuations. The discount rate is applied to the present value of costs and benefits to arrive at the discounted (future) value.

The following calculation clearly illustrates the recommended approach for the comparison of alternatives. The aim is to investigate alternative floor finishes (limited here to carpet and parquetry) and select the material of best value. Only objective matters of total value directly related to each alternative are considered. The presented example has been chosen because it depicts the need to fully consider both capital and operating costs if the most cost-effective solution is to be found. Although parquetry is demonstrated to represent better value than carpet, it should be noted that carpet has the lowest capital cost and both flooring choices are of approximately equal merit when discounting is ignored and the face value of total costs is merely accumulated. No differential income is expected, so income is ignored.

Step 1 Determine the base information for the study

inflation rate (% p.a.): 8
escalation rate for cleaning (% p.a.): 8
escalation rate for energy (% p.a.): n/a
escalation rate for repair (% p.a.): 8
escalation rate for replacement (% p.a.): 8
interest rate on equity (% p.a.): 12
interest rate on borrowing (% p.a.): 17
proportion equity (%): 80
taxation rate on equity (% p.a.): 39
taxation rate on borrowing (% p.a.): 39
affordability rate (% p.a.): −1
period of financial interest (years): 25

Step 2 Determine the base information for each alternative

	Carpet	Parquetry
construction cost ($/m^2)	40.00	110.00
annual cleaning cost ($/m^2)	1.50	1.00
annual energy cost ($/m^2)	nil	nil
annual repair cost ($/m^2)	nil	nil
intermittent repair period (years)	5	10
intermittent repair cost ($/m^2)	5.00	5.00
estimated component life (years)	10	30
replacement cost ($/m^2)	45.00	n/a
asset value growth (% p.a.)	9	9
depreciation allowance (% p.a.)	30 (DV) or 20 (PC)	2.5 (PC)

Step 3 Calculate the expected appreciation rate

$$\text{Asset appreciation rate } (a) = \frac{1+a}{1+f} - 1$$
$$= \frac{1+0.09}{1+0.08} - 1$$
$$= 0.0093$$

Step 4 Calculate the expected investment return

$$\text{Investment return } (i) = (i_1 p_1)(i_2 p_2)$$
$$= [12 \times (1 - 0.39) \times 0.8] + [17 \times (1 - 0.39) \times 0.2]$$
$$= 7.93$$

Step 5 Calculate the expected discount rates

Note that only one discount rate is needed since no differential price level changes are expected:

$$\text{discount rate } (d) = \frac{(1+i)(1+r)}{1+e} - 1$$
$$= \frac{(1+0.0793) \times (1 + (-0.01))}{1 + 0.08} - 1$$
$$= -0.0105$$

Note that depreciation is based on a fixed future cost, and hence must be converted into present value terms before discounting.

The conversion for inflation is: present value (PV) = construction cost $(1 + f)^{-n}$

Step 6 Determine the discounted future value for carpet

capital cost = \$40.00

$$\text{annual cleaning} = \frac{PV(1 - [1 + d]^{-n})}{d}$$

$$= \frac{\$1.50 \times (1 - [1 + (-0.0105)]^{-25})}{-0.0105}$$

$$= \$43.14$$

annual energy = nil

annual repair = nil

intermittent repair

$$= PV(1 + d)^{-n} \text{ for 5, 15 and 25 years}$$
$$= \$5 \times ([1 + (-0.0105)]^{-5} + [1 + (-0.0105)]^{-15} + [1 + (-0.0105)]^{-25})$$
$$= \$17.64$$

$$\text{replacement} = PV(1 + d)^{-n} \text{ for 10 and 20 years}$$
$$= \$45.00 \times ([1 + (-0.0105)]^{-10} + [1 + (-0.0105)]^{-20})$$
$$= \$105.59$$

Step 7 Determine the residual value for carpet

$$\text{value of asset} = PV(1 + a)^n$$
$$= \$40.00 \times (1 + 0.0093)^{25}$$
$$= \$50.42$$

$$\text{residual value (half life expired)} = \frac{\$50.42 \times 5 \times [1 + (-0.0105)]^{-25}}{10}$$

$$= \$32.82$$

$$\text{capital gain (less 5\% selling costs)} = \frac{\$(50.42 - 40.00) \times [1 + (-0.0105)]^{-25}}{1.05}$$

$$= \$12.95$$

Step 8 Determine the discounted future value for parquetry

capital cost = \$110.00

$$\text{annual cleaning} = \frac{PV(1 - [1 + d]^{-n})}{d}$$

$$= \frac{\$1.00 \times (1 - [1 + (-0.0105)]^{-25})}{-0.0105}$$

$$= \$28.76$$

annual energy = nil

annual repair = nil

$$\text{intermittent repair} = PV(1 + d)^{-n} \text{ for 10 and 20 years}$$
$$= \$5 \times ([1 + (-0.0105)]^{-10} + [1 + (-0.0105)]^{-20})$$
$$= \$11.73$$

replacement = nil (replacement is not due until the 30th year)

Step 9 Determine the residual value for parquetry

$$\text{value of asset} = PV(1+a)^n$$
$$= \$110.00 \times (1+0.0093)^{25}$$
$$= 138.64$$

$$\text{residual value (5/6 life expired)} = \frac{\$138.64 \times 5 \times [1+(-0.0105)]^{-25}}{30}$$
$$= \$30.09$$

$$\text{capital gain (less 5\% selling costs)} = \frac{\$(138.64 - 110.00) \times [1+(-0.0105)]^{-25}}{1.05}$$
$$= \$35.52$$

Step 10 Determine the discounted future value for depreciation

Year		Carpet		Parquetry
		DV ($/m^2)	PC ($/m^2)	PC ($/m^2)
1		11.23	7.49	2.57
2		7.36	7.01	2.41
3		4.82	6.55	2.25
4		3.16	6.13	2.11
5		2.07	5.74	1.97
6		1.35	-	1.85
7		0.89	-	1.73
8		0.58	-	1.62
9		0.38	-	1.51
10		0.25	-	1.42
11	carpet replaced >	12.48	8.32	1.32
12		8.17	7.78	1.24
13		5.35	7.28	1.16
14		3.51	6.82	1.09
15		2.30	6.38	1.02
16		1.50	-	0.95
17		0.99	-	0.89
18		0.65	-	0.83
19		0.42	-	0.78
20		0.28	-	0.73
21	carpet replaced >	13.87	9.25	0.68
22		9.08	8.65	0.64
23		5.95	8.10	0.60
24		3.90	7.58	0.56
25		2.54	7.08	0.51
Total:		103.08	110.16	32.44

Step 11 Compare alternatives and select optimum solution

	Carpet ($/m2)		Parquetry ($/m2)	
capital cost		40.00		110.00
operating cost		166.37		40.49
add				
capital gains tax	12.92		35.52	
	x 0.39	= 5.04	x0.39	= 13.85
less				
deductions from income	60.78		40.49	
depreciation	110.16		32.44	
	170.94		72.93	
	x 0.39	= 66.67	x 0.39	= 28.44
less				
residual value		32.82		30.09
Total:		$111.92		$105.81

Given that the original assumptions relating to each alternative are correct, the outcome of the comparison can be reversed by adjusting one or more items of base information, such as selecting a higher interest rate, reassessing the period of financial interest of the owner and using a different tax rate. The previous example is based on a corporate investor within the private sector that is subject to tax at the rate of 39 cents in the dollar. In order to test the sensitivity of the comparison, the calculations can be repeated with a taxation rate equal to zero (e.g. taxation is ignored). The new comparison between carpet and parquetry using a recalculated discount rate of 0.0359 is shown below.

Step 11a Compare alternatives and select optimum solution (tax ignored)

	Carpet ($/m2)		Parquetry ($/m2)	
capital cost		40.00		110.00
operating cost		87.54		22.31
add				
capital gains tax	nil		nil	
	x 0.39	= nil	x0.39	= nil
less				
deductions from income	nil		nil	
depreciation	nil		nil	
	nil		nil	
	x 0.39	= nil		x 0.39 = nil
less				
residual value		10.44		9.57
Total:		$117.10		$122.74

The revised comparison now justifies carpet as the best value for money. Even though carpet has greater taxation benefits than parquetry, the minor change in discount rate is sufficient to cause a reassessment of the optimum design choice. Expenditure in the future becomes less significant, causing the value of operating costs to be effectively halved. The lower capital cost of carpet is thus given more importance. The sensitivity of the discount rate can be tested for values between -10% and +10% for each alternative, both before and after tax. The results of this investigation are presented below:

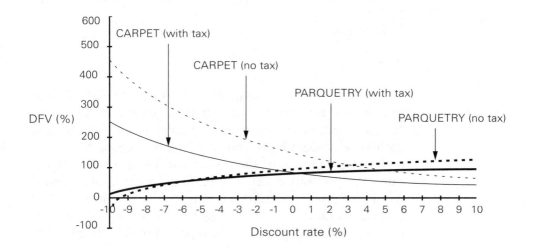

MEASUREMENT OF LIFE-COSTS

It is clear that the discount rate can have a significant impact on the outcome of alternative comparisons. In contrast, present value does not aim to compare but merely to plan and hence does not involve discounting or the calculation of equivalent value. The following calculation illustrates the measurement of life-costs in terms of present value for parquetry.

Step 1 Determine the base information for the study

inflation rate (% p.a.): n/a
period of financial interest (years): 25

Step 2 Determine the base information for parquetry

> construction cost ($/m^2): 110.00
> annual cleaning cost ($/m^2): 1.00
> annual energy cost ($/m^2): nil
> annual repair cost ($/m^2): nil
> intermittent repair period (years): 10
> intermittent repair cost ($/m^2): 5.00
> estimated component life (years): 30
> replacement cost ($/m^2): n/a

Step 3 Determine the present value of parquetry

> capital cost = 110.00
> annual cleaning = $1.00 × 25 years
>   = $25.00
> annual energy = nil
> annual repair = nil
> intermittent repair = $5.00   for 10 and 20 years
>   = $10.00
> replacement = nil

In this example the life-cost of parquetry is calculated as $145.00 compared with a capital cost of $110.00 per square metre, representing an increase of 31.82% over 25 years. Similarly the life-cost of carpet can be shown to equal $182.50 compared to a capital cost of $45.00 per square metre, thus representing an increase of 356.25% over 25 years.

APPENDIX 2

CASE STUDY

DESIGN COST SUMMARY

for

ALBION PARK HIGH SCHOOL

for

DEPARTMENT OF SCHOOL EDUCATION

May, 1989

PROJECT DETAILS

TOTAL PROJECT   ALBION PARK HIGH SCHOOL   Page: 2

TOTAL PROJECT — BUILDING AND EXTERNAL WORKS

PROJECT NAME	Albion Park High School
LOCATION	Church Street, Albion Park, New South Wales
CLIENT	Department of School Education
PERIOD OF FINANCIAL INTEREST	100 years (i.e. indefinite ownership)
CURRENT DATE	May, 1989 INDEX 192.5 (C.P.I.)
DESIGN RISK ALLOWANCE	- %
LOCALITY INDEX	100
MARKET INDEX	100
BUILDING TYPE	New High School (1000 pupil)
GENERAL DESCRIPTION	Generally single storey construction comprising face brick external walls, concrete ground slab, steel roof framing, metal roof cladding and aluminium windows. Block D/E comprising two storey construction with concrete upper floor and columns. External works including covered ways, paving and landscaping.
SPECIAL FACTORS	Bulk earthworks under separate contract.
DRAWING REFERENCES	Architectural (A01-A92) Mechanical (M01-M10)
	Structural (S01-S37) Hydraulics (H01-H35)
	Civil (C01-C04) Landscape (L01-L07)
	Electrical (E01-E36)

TARGET LIFE-COST — Present Value: 36,745,953

PROJECT SUMMARY

	Capital	Operating
Building Cost	6,971,978	21,224,843
Net Project Cost	8,411,620	25,533,826
Gross Project Cost	8,571,810	25,708,826

BUILDING AREAS

				$/m2	Building Cost
Fully Enclosed Covered Area	m²	7,565	77.21 %	3,250.92	24,593,180
Unenclosed Covered Area	m²	2,153	21.97 %	1,625.46	3,499,611
Unenclosed Uncovered Area	m²	80	0.82 %	1,300.37	104,029
GROSS FLOOR AREA	m²	9,798	100.00 %	2,877.81	28,196,821
Useable Floor Area	m²	5,829	-	4,837.33	
Net Rentable Area	m²	n/a	-	n/a	
Building Area	m²	10,162	-	2,774.73	

EFFICIENCY INDICATORS

Area Efficiency	%	59.49
Operating Efficiency	%	30.04
Wall/Floor Area Ratio	:1	
P.O.P. Ratio	%	
Boundary Co-efficient	:1	
Number of Storeys	No	
Floor to Floor Height	m	
Total Building Height	m	

ECONOMIC ASSUMPTIONS

Inflation Rate	%	8.00
Escalation Rate (Cleaning)	%	8.00
Escalation Rate (Energy)	%	10.00
Escalation Rate (Repair)	%	8.00
Escalation Rate (Replacement)	%	8.00
Affordability Rate	%	nil

LIFE-COST SUMMARY

Capital:	
Land	n/a
Construction	8,571,810
Purchase	871,052
Exclusions	1,079,956
Rise and Fall	514,309
Operating:	
Ownership	11,696,051
Maintenance	14,012,775
Occupancy	not included
Exclusions	none
Selling	n/a
Finance:	n/a
TOTAL:	36,745,953

GENERAL SUMMARY

TOTAL PROJECT ALBION PARK HIGH SCHOOL Page: 3

CAPITAL COSTS

	GFA	$/m2	Present Value
Proportion of Preliminaries	-	-	87,029
A Gymnasium and F.S.U.	969 m²	772.90	748,942
B Administration	567 m²	820.03	464,959
C Performing Arts, Music, Home Economics	1,528 m²	755.07	1,153,753
D/E General Learning	3,269 m²	626.45	2,047,859
F Science	880 m²	711.05	625,721
G Industrial Arts	1,088 m²	701.19	762,900
H Arts	768 m²	761.22	584,615
I Library	633 m²	677.77	429,026
J Agricultural Science	96 m²	699.73	67,174
BUILDING COST	9,798 m²	711.57	6,971,978
Centralised Energy Systems	-	-	-
Alterations and Renovations	-	-	-
Site Works		78.83	772,336
External Services		66.27	649,335
External Alterations and Renovations		-	-
Proportion of Preliminaries		1.83	17,971
NET PROJECT COST		858.50	8,411,620
Special Provisions		16.35	160,190
GROSS PROJECT COST		874.85	8,571,810
Land			n/a
Purchase Costs		88.90	871,052
Exclusions		110.22	1,079,956
Rise and Fall		52.49	514,309
Finance Costs			n/a
ANTICIPATED TOTAL COMMITMENT		1,126.47	11,037,127

OPERATING COSTS

	GFA	$/m²	Present Value
Proportion of Preliminaries	-	-	2,285,921
A Gymnasium and F.S.U.	969 m²	2,199.74	2,131,547
B Administration	567 m²	1,823.29	1,033,803
C Performing Arts, Music, Home Economics	1,528 m²	2,092.17	3,196,832
D/E General Learning	3,269 m²	1,547.22	5,057,849
F Science	880 m²	1,904.38	1,675,852
G Industrial Arts	1,088 m²	2,222.80	2,418,405
H Arts	768 m²	1,875.25	1,440,194
I Library	633 m²	2,791.83	1,767,231
J Agricultural Science	96 m²	2,262.59	217,209
BUILDING COST	9,798 m2	2,166.24	21,224,843
Centralised Energy Systems	-	-	-
Alterations and Renovations	-	-	-
Site Works		330.24	3,235,654
External Services		62.18	609,250
External Alterations and Renovations		-	-
Proportion of Preliminaries		47.36	464,079
NET PROJECT COST		2,606.02	25,533,826
Special Provisions		17.86	175,000
GROSS PROJECT COST		2,623.89	25,708,826
Occupancy Costs			not included
Exclusions			none
Selling Costs			n/a
Finance Costs			n/a
ANTICIPATED TOTAL COMMITMENT		2,623.89	25,708,826

GENERAL SUMMARY

ALBION PARK HIGH SCHOOL

TOTAL PROJECT

Page: 4

CAPITAL COSTS

[$ 105,000]

	BC%	$/m²	Present Value
Preliminaries	-	-	-
Substructure	11.58	82.40	807,404
Superstructure	47.94	341.16	3,342,672
Finishes	14.70	104.59	1,024,783
Fittings	5.81	41.33	404,992
Services	18.72	133.20	1,305,098
Proportion of Preliminaries	1.25	8.88	87,029
Design Risk Allowance	-	-	-
Locality Allowance	-	-	-
Market Allowance	-	-	-
BUILDING COST	100.00 %	711.57	6,971,978
Centralised Energy Systems	-	-	-
Alterations and Renovations	-	-	-
Site Works		78.83	772,336
External Services		66.27	649,335
External Alterations and Renovations		-	-
Proportion of Preliminaries		1.83	17,971
NET PROJECT COST		858.50	8,411,620
Special Provisions		16.35	160,190
GROSS PROJECT COST		874.85	8,571,810
Land			n/a
Purchase Costs		88.90	871,052
Exclusions		110.22	1,079,956
Rise and Fall		52.49	514,309
Finance Costs			n/a
ANTICIPATED TOTAL COMMITMENT		1,126.47	11,037,127

OPERATING COSTS

[$ 2,750,000]

	BC%	$/m²	Present Value
Preliminaries	-	-	-
Substructure	0.18	4.00	39,160
Superstructure	17.44	377.70	3,700,689
Finishes	15.81	342.45	3,355,337
Fittings	21.26	460.52	4,512,183
Services	34.54	748.27	7,331,553
Proportion of Preliminaries	10.77	233.30	2,285,921
Design Risk Allowance	-	-	-
Locality Allowance	-	-	-
Market Allowance	-	-	-
BUILDING COST	100.00 %	2,166.24	21,224,843
Centralised Energy Systems	-	-	-
Alterations and Renovations	-	-	-
Site Works		330.24	3,235,654
External Services		62.18	609,250
External Alterations and Renovations		-	-
Proportion of Preliminaries		47.36	464,079
NET PROJECT COST		2,606.02	25,533,826
Special Provisions		17.86	175,000
GROSS PROJECT COST		2,623.89	25,708,826
Occupancy Costs			not included
Exclusions			none
Selling Costs			n/a
Finance Costs			n/a
ANTICIPATED TOTAL COMMITMENT		2,623.89	25,708,826

ELEMENTAL SUMMARY

TOTAL PROJECT

CAPITAL COSTS

		BC %	$/m²	Quantity	Rate	Present Value
SB	SUBSTRUCTURE	11.58	82.40	8,508 m²	94.90	807,404
	SUPERSTRUCTURE					
UF	Upper Floors	3.86	27.47	1,595 m²	168.74	269,146
SC	Staircases	0.35	2.50	54 m²	454.06	24,519
RF	Roof	13.72	97.62	9,113 m²	104.96	956,484
EW	External Walls	13.34	94.94	5,917 m²	157.21	930,207
WW	Windows	7.40	52.69	1,230 m²	419.70	516,235
ED	External Doors	1.92	13.68	254 m²	527.58	134,005
NW	Internal Walls	3.87	27.55	3,900 m²	69.22	269,967
NS	Internal Screens	0.89	6.35	219 m²	283.92	62,178
ND	Internal Doors	2.58	18.36	489 m²	367.96	179,931
	FINISHES					
WF	Wall Finishes	2.08	14.77	13,717 m²	10.55	144,748
FF	Floor Finishes	4.49	31.98	9,348 m²	33.52	313,383
CF	Ceiling Finishes	8.13	57.83	9,255 m²	61.23	566,652
	FITTINGS					
FT	Fitments	3.17	22.54	-	-	220,856
SE	Special Equipment	2.64	18.79	-	-	184,136
	SERVICES					
SF	Sanitary Fixtures	0.71	5.05	183 No	270.41	49,485
PD	Sanitary Plumbing	1.84	13.08	550 Fu	233.05	128,176
WS	Water Supply	2.27	16.17	7,565 m²	20.94	158,385
GS	Gas Service	1.11	7.89	7,469 m²	10.36	77,349
HC	Heating and Cooling	2.95	21.03	7,469 m²	27.58	206,010
AC	Air Conditioning			m²		
FP	Fire Protection	0.22	1.59	7,565 m²	2.06	15,593
ES	Electrical Services	9.61	68.39	7,565 m²	88.58	670,100
TS	Transportation Systems			-	-	
SS	Special Services			-	-	

ALBION PARK HIGH SCHOOL

OPERATING COSTS

		BC %	$/m²	Quantity	Rate	Present Value
SB	SUBSTRUCTURE	0.18	4.00	8,508 m²	4.60	39,160
	SUPERSTRUCTURE					
UF	Upper Floors			1,595 m²		
SC	Staircases	0.27	5.77	54 m²	1046.43	56,507
RF	Roof	4.25	92.00	9,113 m²	98.91	901,392
EW	External Walls	1.75	38.00	5,917 m²	62.92	372,299
WW	Windows	7.47	161.92	1,230 m²	1289.80	1,586,453
ED	External Doors	0.96	20.80	254 m²	802.47	203,828
NW	Internal Walls	0.13	2.87	3,900 m²	7.21	28,138
NS	Internal Screens	0.92	19.85	219 m²	888.24	194,525
ND	Internal Doors	1.68	36.49	489 m²	731.18	357,547
	FINISHES					
WF	Wall Finishes	0.51	10.97	13,717 m²	7.84	107,476
FF	Floor Finishes	13.79	298.72	9,348 m²	313.10	2,926,871
CF	Ceiling Finishes	1.51	32.76	9,255 m²	34.68	320,990
	FITTINGS					
FT	Fitments	15.76	341.45	-	-	3,345,486
SE	Special Equipment	5.50	119.08	-	-	1,166,697
	SERVICES					
SF	Sanitary Fixtures	2.62	56.81	183 No	3041.93	556,673
PD	Sanitary Plumbing	2.10	45.54	550 Fu	811.28	446,202
WS	Water Supply	5.69	123.32	7,565 m²	159.72	1,208,293
GS	Gas Service	2.18	47.31	7,469 m²	62.06	463,500
HC	Heating and Cooling	1.66	35.94	7,469 m²	47.14	352,110
AC	Air Conditioning			m²		
FP	Fire Protection	0.27	5.84	7,565 m²	7.56	57,175
ES	Electrical Services	20.01	433.52	7,565 m²	561.48	4,247,600
TS	Transportation Systems			-	-	
SS	Special Services			-	-	

ELEMENTAL SUMMARY — TOTAL PROJECT

CAPITAL COSTS

		Quantity	Rate	Present Value
CE	CENTRALISED ENERGY SYSTEMS	-	-	-
AR	ALTERATIONS AND RENOVATIONS	-	-	-
	SITE WORKS			
XP	Site Preparation	-	-	51,733
XR	Roads, Paths and Paved Areas	-	-	304,991
XN	Site Walls, Fencing and Gates	-	-	95,504
XB	Outbuildings and Covered Ways	-	-	71,553
XL	Landscaping and Improvements	-	-	248,555
	EXTERNAL SERVICES			
XK	External Stormwater Drainage	-	-	185,120
XD	External Sewer Drainage	-	-	100,349
XW	External Water Supply	-	-	90,219
XG	External Gas	-	-	109,609
XF	External Fire Protection	-	-	81,638
XE	External Electrical Services	-	-	82,400
XS	External Special Services	-	-	-
XX	EXTERNAL ALTERATIONS AND RENOVATIONS	-	-	-
YY	SPECIAL PROVISIONS	-	-	160,190
PR	PROPORTION OF PRELIMINARIES			
	Building Work	-	-	87,029
	External Works	-	-	17,971

ALBION PARK HIGH SCHOOL — Page: 6

OPERATING COSTS

		Quantity	Rate	Present Value
CE	CENTRALISED ENERGY SYSTEMS	-	-	-
AR	ALTERATIONS AND RENOVATIONS	-	-	-
	SITE WORKS			
XP	Site Preparation	-	-	15,800
XR	Roads, Paths and Paved Areas	-	-	546,006
XN	Site Walls, Fencing and Gates	-	-	223,720
XB	Outbuildings and Covered Ways	-	-	69,986
XL	Landscaping and Improvements	-	-	2,380,142
	EXTERNAL SERVICES			
XK	External Stormwater Drainage	-	-	68,000
XD	External Sewer Drainage	-	-	97,350
XW	External Water Supply	-	-	275,000
XG	External Gas	-	-	1,000
XF	External Fire Protection	-	-	10,000
XE	External Electrical Services	-	-	157,900
XS	External Special Services	-	-	-
XX	EXTERNAL ALTERATIONS AND RENOVATIONS	-	-	-
YY	SPECIAL PROVISIONS	-	-	175,000
PR	PROPORTION OF PRELIMINARIES			
	Building Work	-	-	2,285,921
	External Works	-	-	464,079

OPERATING COST SUMMARY

TOTAL PROJECT   ALBION PARK HIGH SCHOOL   Page: 7

OPERATING COSTS	OWNERSHIP COSTS						MAINTENANCE COSTS				TOTAL
	Cleaning	%	Energy	%	Other	%	Repair	%	Replacement	%	Operating
SB SUBSTRUCTURE							39,160	100.00			39,160
SUPERSTRUCTURE											
UF Upper Floors											
SC Staircases	12,350	21.86					3,420	6.05	40,737	72.09	56,507
RF Roof	70,625	7.84					11,508	1.28	819,259	90.89	901,392
EW External Walls	9,500	2.55					165,770	44.53	197,029	52.92	372,299
WW Windows	174,500	11.00					101,880	6.42	1,310,073	82.58	1,586,453
ED External Doors							68,992	33.85	134,836	66.15	203,828
NW Internal Walls							28,138	100.00			28,138
NS Internal Screens	40,200	20.67							154,325	79.33	194,525
ND Internal Doors							109,445	30.73	246,688	69.27	356,133
FINISHES											
WF Wall Finishes	30,000	27.91					50,920	47.38	26,556	24.71	107,476
FF Floor Finishes	1,340,976	45.82					73,836	2.52	1,512,059	51.66	2,926,871
CF Ceiling Finishes							318,040	98.65	4,364	1.35	322,404
FITTINGS											
FT Fitments	70,400	2.10					5,423	0.16	3,269,663	97.73	3,345,486
SE Special Equipment							341,904	29.31	824,793	70.69	1,166,697
SERVICES											
SF Sanitary Fixtures	372,500	66.92					3,486	0.78	184,173	33.08	556,673
PD Sanitary Plumbing	58,500	13.11			366,000	82.03	24,900	2.06	18,216	4.08	446,202
WS Water Supply			980,000	81.11					203,393	16.83	1,208,293
GS Gas Service			463,500	100.00							463,500
HC Heating and Cooling									352,110	100.00	352,110
AC Air Conditioning											
FP Fire Protection											
ES Electrical Services							43,000	75.21	14,175	24.79	57,175
TS Transportation Systems											
SS Special Services	3,947,600	92.94			300,000	7.06					4,247,600

OPERATING COST SUMMARY

TOTAL PROJECT   ALBION PARK HIGH SCHOOL   Page: 8

OPERATING COSTS	OWNERSHIP COSTS						MAINTENANCE COSTS				TOTAL
	Cleaning	%	Energy	%	Other	%	Repair	%	Replacement	%	Operating
CE CENTRALISED ENERGY SYSTEMS											
AR ALTERATIONS AND RENOVATIONS											
SITE WORKS											
XP Site Preparation							15,800	100.00			15,800
XR Roads, Paths and Paved Areas	287,550	52.66					87,341	16.00	171,115	31.34	546,006
XN Site Walls, Fencing and Gates							16,920	7.56	206,800	92.44	223,720
XB Outbuildings and Covered Ways	21,950	31.36					3,759	5.37	44,277	63.27	69,986
XL Landscaping and Improvements							2,199,125	92.39	181,017	7.61	2,380,142
EXTERNAL SERVICES											
XK External Stormwater Drainage	68,000	100.00									68,000
XD External Sewer Drainage	24,000	24.65					73,350	75.35			97,350
XW External Water Supply			275,000	100.00							275,000
XG External Gas							1,000	100.00			1,000
XF External Fire Protection							10,000	100.00			10,000
XE External Electrical Services			157,900	100.00							157,900
XS External Special Services											
XX EXTERNAL ALTERATIONS AND RENOVATIONS											
YY SPECIAL PROVISIONS	125,000	71.43					50,000	28.57			175,000
PR PROPORTION OF PRELIMINARIES											
Building Work	415,622	18.18			1,662,488	72.73	207,811	9.09			2,285,921
External Works	84,378	18.18			337,512	72.73	42,189	9.09			464,079
OPERATING COST TOTALS:	3,206,051	12.47	5,824,000	22.65	2,666,000	10.37	4,097,117	15.94	9,915,658	38.57	25,708,826

CASH FLOW SUMMARY

TOTAL PROJECT   ALBION PARK HIGH SCHOOL

PRESENT VALUE

Total Current Operating Expenditure

Year	Value	Year	Value	Year	Value
1991	168,727	2024	168,727	2057	168,727
1992	168,727	2025	208,993	2058	208,993
1993	168,727	2026	168,727	2059	168,727
1994	168,727	2027	168,727	2060	168,727
1995	208,992	2028	168,727	2061	168,727
1996	168,727	2029	168,727	2062	168,727
1997	168,727	2030	1,251,335	2063	1,251,335
1998	168,727	2031	168,727	2064	168,727
1999	168,727	2032	168,727	2065	168,727
2000	292,293	2033	168,727	2066	168,727
2001	168,727	2034	168,727	2067	168,727
2002	168,727	2035	411,547	2068	411,547
2003	168,727	2036	168,727	2069	168,727
2004	168,727	2037	168,727	2070	1,292,499
2005	411,547	2038	168,727	2071	168,727
2006	168,727	2039	168,727	2072	168,727
2007	168,727	2040	564,795	2073	564,795
2008	168,727	2041	168,727	2074	168,727
2009	168,727	2042	168,727	2075	168,727
2010	555,846	2043	168,727	2076	168,727
2011	168,727	2044	168,727	2077	168,727
2012	168,727	2045	208,993	2078	208,993
2013	168,727	2046	168,727	2079	168,727
2014	168,727	2047	168,727	2080	1,307,084
2015	333,948	2048	168,727	2081	168,727
2016	168,727	2049	168,727	2082	168,727
2017	168,727	2050	1,570,637	2083	1,570,637
2018	168,727	2051	168,727	2084	168,727
2019	168,727	2052	168,727	2085	208,993
2020	1,307,084	2053	168,727	2086	168,727
2021	168,727	2054	168,727	2087	168,727
2022	168,727	2055	208,993	2088	208,993
2023	168,727	2056	168,727	2089	168,727

FUTURE VALUE

Affordability-adjusted Operating Expenditure

Year	Value	Year	Value	Year	Value
1991	169,805	2024	219,171	2057	309,618
1992	170,903	2025	261,450	2058	313,306
1993	172,022	2026	223,233	2059	317,062
1994	173,161	2027	225,321	2060	444,454
1995	214,588	2028	227,448	2061	324,784
1996	175,504	2029	229,614	2062	328,752
1997	176,708	2030	1,314,428	2063	332,794
1998	177,935	2031	234,067	2064	336,911
1999	179,184	2032	236,355	2065	708,879
2000	304,023	2033	238,686	2066	345,375
2001	181,752	2034	241,060	2067	349,724
2002	183,071	2035	486,298	2068	354,155
2003	184,415	2036	245,941	2069	358,667
2004	185,786	2037	248,450	2070	1,487,035
2005	430,000	2038	251,004	2071	367,944
2006	188,600	2039	253,607	2072	372,712
2007	190,047	2040	652,325	2073	377,568
2008	191,520	2041	258,956	2074	382,514
2009	193,021	2042	261,706	2075	427,817
2010	581,668	2043	264,506	2076	392,682
2011	196,106	2044	267,358	2077	397,908
2012	197,691	2045	310,529	2078	403,230
2013	199,306	2046	273,222	2079	408,652
2014	200,951	2047	276,236	2080	1,552,530
2015	367,847	2048	279,305	2081	419,797
2016	204,333	2049	282,431	2082	425,525
2017	206,070	2050	1,687,526	2083	431,359
2018	207,840	2051	288,859	2084	437,301
2019	209,643	2052	292,162	2085	483,619
2020	1,349,837	2053	295,526	2086	449,517
2021	213,350	2054	298,953	2087	455,796
2022	215,255	2055	342,709	2088	462,190
2023	217,195	2056	305,998	2089	468,703

Present Value Life-Costs by Type

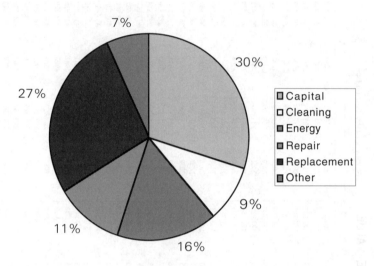

Present Value Life-Costs by Element Grouping

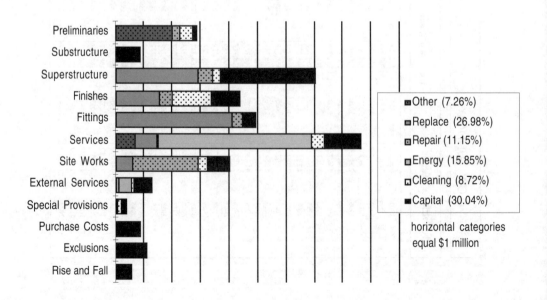

Present Value Capital Costs by Function

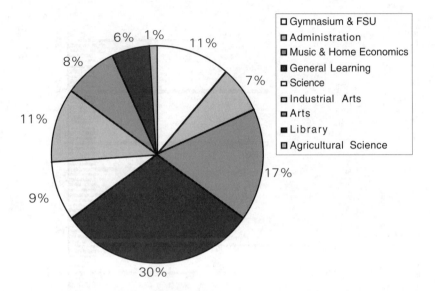

Present Value Operating Costs by Function

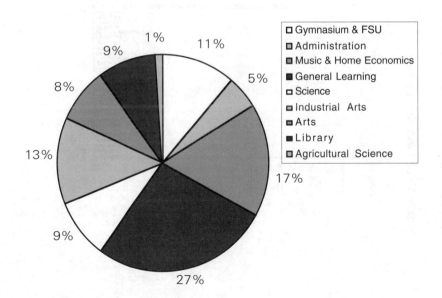

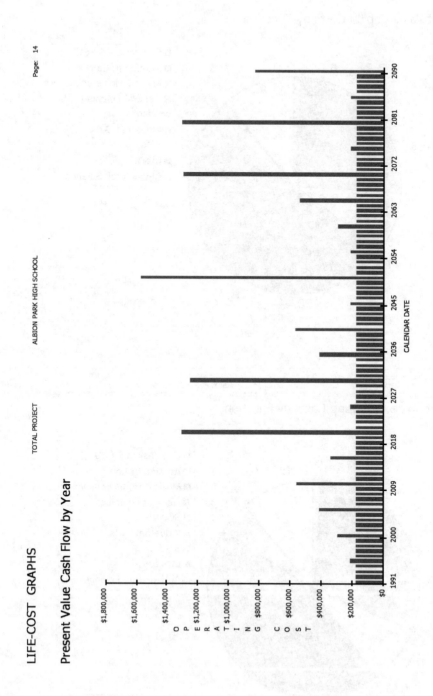

LIFE-COST GRAPHS

TOTAL PROJECT

ALBION PARK HIGH SCHOOL

Page: 14

Present Value Cash Flow by Year

APPENDIX 3

SAMPLE EXAMINATION PAPERS

PAPER 1

Time allowed is 3 hours.

Question 1 tests knowledge and comprehension, Questions 2, 3 and 4 test application and analysis, and Question 5 tests synthesis and evaluation.

All questions are to be attempted. Marks are indicated against each question.

Calculators may be used. Otherwise this is a closed-book exam.

Question 1   (5 marks each - 50 marks total)

Provide short answers to each of the following questions:

(a) What is the discounted value in 10 years' time of $1 000 today at a discount rate of 5% per annum?

(b) What is the definition of internal rate of return?

(c) What is the primary purpose of using risk analysis techniques as part of a life-cost comparison?

(d) What are the principal benefits of post-occupancy evaluation?

(e) What is the difference in meaning between intragenerational and intergenerational equity?

(f) What is the essence of the argument supporting the view that discounting does not disadvantage future generations?

(g) What are full-year effect costs and why should they be used with caution?

(h) What are types of functional plans used in the management of large facilities?

(i) Should interest payments on borrowed capital be included in either a discounted cash flow analysis or a life-cost comparison? Why?

(j) What are denial-of-use costs?

Question 2   (10 marks)
Briefly describe the ways in which a building/facility can become obsolete.

Question 3   (10 marks)
Discount rates are suggested as a combination of project-related, product-related and investor-related considerations. How would you justify this statement?

Question 4   (10 marks)
Compare the cost-effectiveness of two alternatives, designated as Option A and Option B, given the following information:

time horizon for study: 10 years

test discount rate: 3% per annum real

	Option A	Option B
capital cost	$1000	$5000
annual operating cost	$500	$150
extra maintenance cost at Year 5	$300	$500
residual value	$250	$1000

Assuming taxation issues are ignored, which option would you recommend? Your answer should include a sensitivity analysis on the discount rate using 0% and 6% as best-case and worse-case scenarios.

Question 5   (20 marks)
Answer <u>one</u> of the following three parts:

(a) Value management is merely organised common sense. Provide the case FOR and AGAINST this proposition, and your opinion.

(b) Cost-benefit analysis is an inherently flawed technique because it uses money as a universal metric. Provide the case FOR and AGAINST this proposition, and your opinion.

(c) Sustainable development is an interesting area of study but not one that has much relevance to the professional practice of building economists (quantity surveyors). Provide the case FOR and AGAINST this proposition, and your opinion.

PAPER 2

Time allowed is 3 hours.

Question 1 tests knowledge and comprehension, Questions 2, 3 and 4 test application and analysis, and Question 5 tests synthesis and evaluation.

All questions are to be attempted. Marks are indicated against each question.

Calculators may be used. Otherwise this is a closed-book exam.

Question 1 (5 marks each - 50 marks total)

Provide short answers to each of the following questions:

(a) Why is the life-costing technique attracting renewed interest from the construction industry?

(b) What is the purpose of calculating discounted present value?

(c) What is the basis of the time preference philosophy of discounting?

(d) Why is there a potential conflict of interest when choosing whether to base life-cost studies on economic life or the period of financial interest of the client?

(e) Given you had access to historical data for the operation of existing buildings, why would you still view these data with caution when attempting to apply them to a new project?

(f) What is the formula for calculating the rate of change in worth of an asset? Identify each variable.

(g) What are the differences between traditional discounted cash flow analysis and life-cost comparisons?

(h) What is the link between post-occupancy evaluation and worker productivity?

(i) Give three different examples of how building design decisions can reduce occupancy costs.

(j) What advantages can information technology offer to facility management processes?

Question 2 (10 marks)

You have signed a binding agreement to defer payment of $1000 for two years at no interest (i.e. future cost = $1000). Calculate the present value, discounted present value and discounted future value for this transaction assuming annual inflation and product escalation at 3% per annum, investment return (based on 100% use of equity) at 10% per annum, taxation at 36% and a predicted rate of affordability change of -1% per annum.

Question 3 (10 marks)

Calculate and compare the cost-effectiveness of two alternatives, designated as Option A and Option B, given the following information:

time horizon for study: 12 years
test discount rate: 2% per annum real

	Option A	Option B
capital cost	$5 000	$12 000
annual operating cost	$1 000	$500
extra maintenance cost at Yr 5 & 10	$500	$1 000
residual value	$1 000	$3 000

Assuming taxation issues are ignored, which option would you recommend? Your answer should include a sensitivity analysis on the discount rate using 0% and 4% as best-case and worse-case scenarios.

Question 4 (10 marks)

Briefly describe the steps involved in the life-cost planning and analysis of a project and the types of documents that would be produced to support the cost management process.

Question 5 (20 marks)

Answer <u>one</u> of the following three parts:

(a) There is no such thing as sustainable development. Provide the case FOR and AGAINST this proposition, and your opinion.

(b) If we had a historical database of building material and system performances, life-cost studies would be able to be undertaken more easily and with greater reliability. Provide the case FOR and AGAINST this proposition, and your opinion.

(c) Discounting is a process that works against intergenerational equity, and hence is in conflict with sustainable development goals. Provide the case FOR and AGAINST this proposition, and your opinion.

PAPER 3

Time allowed is 3 hours.

Question 1 tests knowledge and comprehension, Questions 2, 3 and 4 test application and analysis, and Question 5 tests synthesis and evaluation.

All questions are to be attempted. Marks are indicated against each question.

Calculators may be used. Otherwise this is a closed-book exam.

Question 1 (5 marks each - 50 marks total)

Provide short answers to each of the following questions:

(a) How is normal preference for revenue received early and expenditure paid late incorporated into project evaluation decisions?

(b) Why are investment return and inflation opposing factors in the establishment of a real discount rate?

(c) Why are real cash flows and real discount rates preferred over inflated costs and nominal discount rates?

(d) What is the conflict of interest that exists in choosing a time horizon for a life-cost comparison?

(e) What are 'conservatism factors' and how are they applied to risk analysis investigations?

(f) What is embodied energy and why is it of interest to an analysis of sustainability performance?

(g) What does the process of energy auditing have in common with post-occupancy evaluation?

(h) What are two key areas of staff activity that building design can affect and which, if minimised, will lead to substantial savings in occupancy cost?

(i) How does the construction industry achieve a 'systems approach' to value management when buildings are generally too complex to be considered in a holistic manner?

(j) For what reasons do you think environmental issues will drive building economics (quantity surveying) activities more towards assessment of operational performance of buildings?

Question 2 (10 marks)

Given an inflation-linked asset having a present cost of $1000 in 1999, calculate its present value, future cost, future value, discounted present value and discounted future value for the year 2009. Assume the weighted investment return is 10% per annum, general inflation is 5% per annum, affordability is 1% per annum, no expected differential price level changes and no taxation.

Question 3 (10 marks)

What is the argument for inclusion of an affordability rate in life-cost comparisons?

Question 4 (10 marks)

The following cash flow represents two competing projects that you have been asked to evaluate:

	Project A		Project B	
Year	Cost	Benefit	Cost	Benefit
0	5 000 000	0	5 000 000	0
1	10 000 000	0	12 000 000	0
2	500 000	2 000 000	850 000	2 500 000
3	500 000	2 000 000	850 000	2 500 000
4	500 000	2 000 000	850 000	2 500 000
5	1 000 000	2 000 000	1 700 000	2 500 000
6	500 000	2 000 000	850 000	2 500 000
7	500 000	2 000 000	850 000	2 500 000
8	500 000	2 000 000	850 000	2 500 000
9	500 000	2 000 000	850 000	2 500 000
10	1 000 000	20 750 000	1 700 000	23 125 000

Using a discount rate of 5%, what is the BCR for each project and which project would you judge as the better investment?

Question 5 (20 marks)

Answer <u>one</u> of the following three parts:

(a) Life-cost studies will struggle to be a major client service, since no matter how important the concept they can never hope to 'predict the unpredictable'. Provide the case FOR and AGAINST this proposition, and your opinion.

(b) The failure to achieve significant and widespread advances in the design of sustainable buildings can be blamed primarily on the engagement of 'cost-fixated' quantity surveyors. Provide the case FOR and AGAINST this proposition, and your opinion.

(c) Discounting is a process that works against intergenerational equity, and hence is in conflict with sustainable development goals. Provide the case FOR and AGAINST this proposition, and your opinion.

TUTORIAL WORKED EXAMPLES

TUTORIAL CHAPTER 1

Using Microsoft Excel, the world oil price data provided in Table 1.1 can be plotted as an XY scatter chart with a linear regression trend line and equation added (see illustration below). The regression equation indicates that the value for 2005 is 23.08 US$/barrel. However, the corresponding value for R^2 indicates that the regression line is not a strong fit for this data series.

Note that R^2 values vary between 0 (no relationship) and 1 (perfect relationship). An R^2 value of 0.2754 (27%) is weak. An R^2 value greater than 70% might be considered strong.

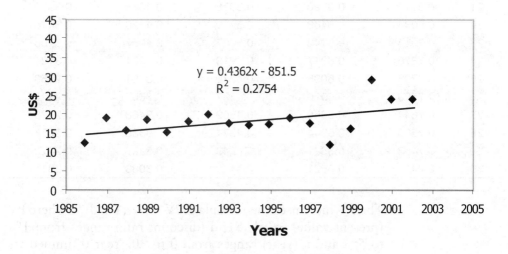

$$y = 0.4362x - 851.5$$
$$R^2 = 0.2754$$

TUTORIAL CHAPTER 2

The PV table based on equivalent values of $1 over 30 years for discount rates between 1% and 5% inclusive is shown below.

Year	1%	2%	3%	4%	5%
0	1.0000	1.0000	1.0000	1.0000	1.0000
1	0.9901	0.9804	0.9709	0.9615	0.9524
2	0.9803	0.9612	0.9426	0.9246	0.9070
3	0.9706	0.9423	0.9151	0.8890	0.8638
4	0.9610	0.9238	0.8885	0.8548	0.8227
5	0.9515	0.9057	0.8626	0.8219	0.7835
6	0.9420	0.8880	0.8375	0.7903	0.7462
7	0.9327	0.8706	0.8131	0.7599	0.7107
8	0.9235	0.8535	0.7894	0.7307	0.6768
9	0.9143	0.8368	0.7664	0.7026	0.6446
10	0.9053	0.8203	0.7441	0.6756	0.6139
11	0.8963	0.8043	0.7224	0.6496	0.5847
12	0.8874	0.7885	0.7014	0.6246	0.5568
13	0.8787	0.7730	0.6810	0.6006	0.5303
14	0.8700	0.7579	0.6611	0.5775	0.5051
15	0.8613	0.7430	0.6419	0.5553	0.4810
16	0.8528	0.7284	0.6232	0.5339	0.4581
17	0.8444	0.7142	0.6050	0.5134	0.4363
18	0.8360	0.7002	0.5874	0.4936	0.4155
19	0.8277	0.6864	0.5703	0.4746	0.3957
20	0.8195	0.6730	0.5537	0.4564	0.3769
21	0.8114	0.6598	0.5375	0.4388	0.3589
22	0.8034	0.6468	0.5219	0.4220	0.3418
23	0.7954	0.6342	0.5067	0.4057	0.3256
24	0.7876	0.6217	0.4919	0.3901	0.3101
25	0.7798	0.6095	0.4776	0.3751	0.2953
26	0.7720	0.5976	0.4637	0.3607	0.2812
27	0.7644	0.5859	0.4502	0.3468	0.2678
28	0.7568	0.5744	0.4371	0.3335	0.2551
29	0.7493	0.5631	0.4243	0.3207	0.2429
30	0.7419	0.5521	0.4120	0.3083	0.2314

The PV table uses the formula $DPV = PV (1 + d)^{-n}$, where PV (present value) equals $1, d (discount rate) ranges from 1% to 5% and n (year) ranges from 0 to 30. Year 0 (immediate payment or receipt) remains unadjusted.

Displaying these data in a graphical form illustrates the compounding effect of the discount rate on future costs. The value of $1 at a 5% discount rate after 30 years is 23.14 cents.

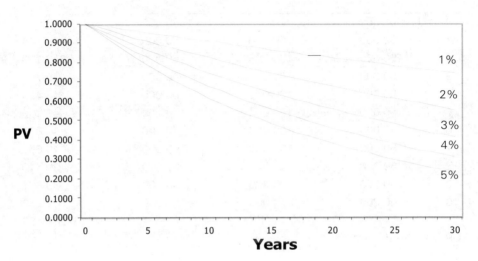

TUTORIAL CHAPTER 3

The following discounted cash flows relate to each investment option.

Option 1

Year	Cost ($)	Benefit ($)	Net benefit ($)	Discount Net Benefit ($)
0	100 000	0	-100 000	-100 000
1	10 000	20 000	10 000	9 709
2	10 000	20 000	10 000	9 426
3	10 000	20 000	10 000	9 151
4	10 000	20 000	10 000	8 885
5	10 000	20 000	10 000	8 626
6	10 000	20 000	10 000	8 375
7	10 000	20 000	10 000	8 131
8	10 000	20 000	10 000	7 894
9	10 000	20 000	10 000	7 664
10	10 000	120 000	110 000	81 850
				NPV = 59 711

BCR = 1.32   IRR = 10%

Note that BCR is calculated as the sum of the discounted benefits divided by the sum of the discounted costs, and IRR is found by trial and error or graphical interpretation as the discount rate that leads to a net present value of zero.

Option 2

Year	Cost ($)	Benefit ($)	Net benefit ($)	Discount Net Benefit ($)
0	500 000	0	-500 000	-500 000
1	50 000	100 000	50 000	48 544
2	50 000	100 000	50 000	47 130
3	50 000	100 000	50 000	45 757
4	50 000	100 000	50 000	44 424
5	50 000	100 000	50 000	43 130
6	50 000	100 000	50 000	41 874
7	50 000	100 000	50 000	40 655
8	50 000	100 000	50 000	39 470
9	50 000	100 000	50 000	38 321
10	50 000	600 000	550 000	409 252
				NPV = 298 557

BCR = 1.32   IRR = 10%

Option 3

Year	Cost ($)	Benefit ($)	Net benefit ($)	Discount Net Benefit ($)
0	250 000	0	-250 000	-250 000
1	10 000	50 000	40 000	38 835
2	10 000	50 000	40 000	37 704
3	10 000	50 000	40 000	36 606
4	10 000	50 000	40 000	35 539
5	10 000	50 000	40 000	34 504
6	10 000	50 000	40 000	33 499
7	10 000	50 000	40 000	32 524
8	10 000	50 000	40 000	31 576
9	10 000	50 000	40 000	30 657
10	10 000	300 000	290 000	215 787
				NPV = 277 232

BCR = 1.83   IRR = 16%

All investments make a positive NPV at 3% discount and therefore are acceptable. Option 2 has the highest NPV. However, Option 3 has a similar NPV but significantly higher BCR and IRR, and therefore would be the best option.

TUTORIAL CHAPTER 4

The following cost plan shows the playground costs over a 30-year time horizon:

30-YEAR LIFE-COST PLAN:

Capital cost		50 000
Operating cost:		
Lawn mowing and watering	39 000	
Lighting	18 000	
Playground repair	15 000	
Repainting line marking	6 000	
Resurfacing pavement	30 000	
Replacing seating/fitments	15 000	123 000
Total:		173 000

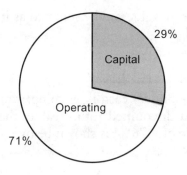

TUTORIAL CHAPTER 5

The following table calculates the actual cost expressed in present value terms:

Year	Plan (PV)	Actual (FC)	BPI	Actual (PV)
0	50 000	48 785	100.0	*48 785
1	2 400	2 505	102.1	2 453
2	2 400	2 490	102.7	2 425
3	2 400	2 500	103.6	2 413
4	2 400	2 610	104.0	2 510
5	3 400	3 560	105.2	*3 384
6	2 400	2 650	106.9	2 479
7	2 400	2 700	108.1	2 498
8	2 400	2 700	110.3	2 448
9	2 400	2 750	110.2	2 495
10	18 400	20 500	111.8	*18 336
	91 000			*90 226

*indicates the project is just under budget

TUTORIAL CHAPTER 6

Future value over 10 years is calculated based on the values for change in worth as shown below:

	PV	w	total FV
purchase	38 000		38 000
fuel costs (annual)	1 500	-0.0285	17 647
maintenance costs (annual)	3 000	-0.0100	31 718
new tyres (every 5 years)	600	-0.0100	1 294
new radiator (every 7 years)	250	-0.0100	268
new exhaust system (every 10 years)	500	-0.0100	553
resale after 10 years	-5 000	-0.0100	-5 529
			83 952

Future value can be summed as it is an equivalent present-day value.

TUTORIAL CHAPTER 7

The annual cash flow comprising present value, future value and discounted future value, the latter based on a discount rate of 4.76%, is shown below:

Year	PV	FV	DFV
0	38 000	38 000	38 000
1	4 500	4 574	4 366
2	4 500	4 650	4 237
3	4 500	4 728	4 112
4	4 500	4 807	3 991
5	5 100	5 519	4 374
6	4 500	4 971	3 760
7	4 750	5 323	3 844
8	4 500	5 142	3 544
9	4 500	5 230	3 441
10	600	1 008	633
	79 950	83 952	74 301

Note that discounted future value has no meaning unless it is used to compare this scenario with one or more alternatives.

The calculations clearly indicate that the cost of running the car (in equivalent dollars) is expected to be more in future years based on escalating fuel prices and a negative affordability rate (i.e. things becoming harder to afford).

TUTORIAL CHAPTER 8

The relationship between total life-cost and total energy is shown in the following chart. The R^2 value of 0.7004 (70%) indicates quite a strong correlation. The regression equation can be used to predict energy values (GJ/m^2) based on life-cost for future projects ($/m^2).

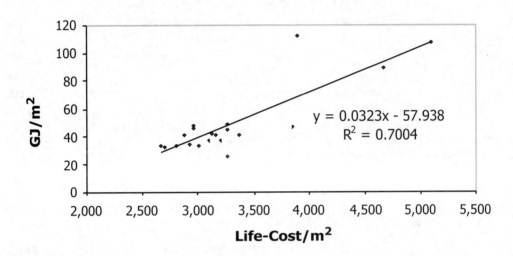

TUTORIAL CHAPTER 9

The following payback periods are derived:

$$\text{Payback period (operating)} = \frac{4\ 500\ 000}{800\ 000 \times 0.25}$$

$$= 22.5 \ \text{years}$$

$$\text{Payback period (sales)} = \frac{4\ 500\ 000}{3\ 500\ 000 - 2\ 100\ 000}$$

$$= 3.21 \ \text{years}$$

$$\text{Payback period (combined)} = \frac{4\ 500\ 000}{200\ 000 + 1\ 400\ 000}$$

$$= 2.81 \ \text{years}$$

TUTORIAL CHAPTER 10

The appropriate methodology for estimating stormwater pollution to a local waterway is the cost of preventing the damage in the first instance. A possible approach is shown below:

Item	Estimate	Total ($)
construct retention pits	20 pits @ $25 000/pit	500 000
clean pits (annual)	20 pits @ $500/pit/p.a	
	=10 000 p.a	
	over 5 years	50 000
		550 000
public education campaign		500 000
	total over 5 years	= 1 050 000

Therefore the annual cost is $210 000 or $87.5/Ml of waste.

TUTORIAL CHAPTER 11

The calculation of value for money is determined as value score divided by life-cost. The higher the index the better:

	value score (index)	life-cost ($/m²)	value for money
Option 1: residential apartments	317	3876	0.0818
Option 2: retail	250	3088	0.0810
Option 3: hotel	295	3502	0.0842

Option 3 is shown to represent the best value.

This simple comparison ignores a number of significant issues:

- *Income* Life-cost considers expenditure flows, and in this case should include differential revenue compared to a base option displaying the lowest return. Alternatively, BCR could be calculated in lieu of life-cost and value for money determined as value score x BCR.
- *Market survey* The assessment of options should consider whether a demand exists for this type of development in the marketplace.
- *Sustainability* The value score would need to account for environmental and social factors that may apply.
- *Risk* How sensitive is each value score and life-cost to change?

TUTORIAL CHAPTER 12

The benchmarked criteria for each option are shown below (values shown as ~~strike-through~~ do not reach set thresholds):

Option	Wealth	Utility	Resources	Impact
A	141	88	77	24
B	329	53	92	42
C	115	69	92	45
D	~~87~~	75	62	48
E	261	73	85	30
F	420	~~33~~	~~123~~	~~88~~

Option	Sustainability Index		
	equal weight	economic bias	social bias
A	5.44	2.28	3.17 (best)
B	4.80	2.98 (best)	1.82
C	2.76	1.31	1.45
D	2.95	1.44	1.51
E	5.48 (best)	2.91	2.57
F	3.78	2.65	

From this analysis Options D and F do not meet required performance benchmarks, and must be dropped or redesigned. When all criteria are given equal weight, Option E is recommended. Should the client have more interest in economic outcomes, Option B is slightly preferred, whereas if social outcomes are important, Option A is preferred. All calculated sustainability indices are greater than 1, and therefore represent net socio-economic gain.

GLOSSARY
OF TERMS

activity report statement of actual performance (including cost, durability, maintenance demands and the like) from construction and/or occupation in a form that facilitates both control and information reuse

affordability changes in the purchasing power of money over time (see *change in worth*)

affordability index indicative measure of changes in affordability through analysis of data on income and expenditure patterns

affordability rate the rate of change in affordability over time as determined by an affordability index

annual equivalent the present value of a series of discounted cash streams expressed as a constant annual amount

annuity a recurring annual payment or receipt

appreciation rate nominal rate at which assets increase in value over time per annum

benchmark a reference point to best practice or performance threshold

bill of quantities document that details the items of work involved in the construction of the project for the purposes of tendering and contract administration

budget overall control mechanism and cost limit for the project based on historical elemental costs from similar projects

capital budgeting a variety of procedures and techniques that ultimately aim to rank and select investment projects based on profitability

capital cost initial acquisition cost of the land and building

capital gains tax tax on the inflation-adjusted profit (after selling expenses) received on sale of an asset, assuming that a profit was made, often provided that the property is not the owner's principal place of residence

capital productivity a philosophy of discounting based on the theoretical earning capacity (opportunity) of money over time

cash flow summary of the inflows and outflows of cash for a project by time period

change in worth the rate of change in the base value of goods and services due to differential price level changes and the diminishing marginal utility (see *escalation* and *affordability*)

cleaning cost expenditure on regular cleaning of building surfaces and components (see *ownership cost*)

comparative value representation of worth in equivalent dollars for the purpose of comparison of two or more alternatives with regard to the changing time value of money

compound interest the effective rate of increase of an investment where the interest is reinvested

comparison evaluation of life-costs for two or more design alternatives (see *discounted future value*)

construction cost cost of labour, material and plant involved in the creation of the building and other improvements to the land, including all supervision, profit and rise and fall during the construction period

cost-benefit analysis technique for the comparison of tangible and intangible costs and benefits over time for alternative projects based on either a social or economic approach

cost-in-use see *life-cost*

cost check process of checking and reporting the estimated cost of each section or element of the project as detailed designs are developed against the cost target given in the life-cost plan

cost control overall process of budgeting, optimising, documenting, monitoring and managing the life-cost of a building project, commencing with the decision to build and concluding when the cost implications of the project are no longer of concern to the owner

denial-of-use cost extra costs or income lost because occupancy or production is delayed as a result of a design decision

depreciation taxation concession for the reduction in value of capital assets over time

differential price level change net increase or decrease in the value of goods and services due to differences in the rate of change between escalation and inflation

differential revenue amount of increase in income between one design alternative and another

diminishing marginal utility economic condition indicating that each additional unit of consumption (or income) is valued less than the previous one (see *marginal utility*)

discount rate rate of adjustment used to account for expected changes in affordability and/or opportunity over time (see *discounting*)

discounted cash flow analysis technique for assessing the return on capital employed in an investment project over its economic life or suitable time horizon, with a view to prioritising alternative courses of action that exceed established profitability thresholds

discounted future value opportunity-adjusted future value – it is used for the comparison of design alternatives and in discounted cash flow analysis calculations

discounted present value opportunity-adjusted present value – ignores changes in value base worth and is thus an oversimplification of the time value of money (not recommended for use)

discounting theoretical adjustment for the time value of money for use when making comparisons between alternatives exhibiting differential timing in the payment and receipt of cash

economic life period of time during which investment in an asset is the least-cost alternative for satisfying a particular objective

elasticity of marginal ultility the percentage change in marginal utility resulting from a 1% increase in consumption (or income)

element portion of a project that fulfils a particular physical purpose irrespective of construction and/or specification

elemental cost analysis collection and compilation of elemental or sub-elemental costs from actual information for the purpose of analysis and reuse

elemental cost planning technique of cost control involving the division of the project into element and sub-element groupings to enhance control and facilitate comparison with similar elements and sub-elements in other buildings

energy cost expenditure on items of energy supply, including electricity, water, gas, oil and the like (see *ownership cost*)

equivalent value comparative worth of future costs and benefits expressed in today's terms

escalation changes in the cost of specific commodities over time (see *change in worth*)

escalation rate the rate of increase in the specific price of goods and services over time, normally divided into cleaning, energy, repair, replacement and other categories for building projects

exchange rate rate of change in the comparative value of money over time or between countries

finance cost expenditure relating to the interest component of loan repayments, establishment and account fees, holding charges and other liabilities associated directly with borrowed capital

future cost actual or 'cheque-book' cost incurred at some future time – it is expressed in current dollars and cannot be accumulated without adjustment for inflation (future cost may be fixed or inflation-linked)

future value comparative value of costs and benefits to future generations determined after adjustment of present value for changes in worth (see *time preference*)

impatience myopic human trait of preference for present control over money and other resources compared to future control

inflation an economic term that describes the general increase in the price of goods and services over time caused by the relative balance between supply and demand

inflation rate the rate of increase in the general price of goods and services over time

intangibles costs and benefits pertaining to social amenity and welfare that are difficult to quantify in monetary terms (see *cost-benefit analysis*)

intergenerational equity the concept that future generations should be no worse off than present generations due to the decisions made by the latter

internal rate of return the discount rate that leads to a net present value of zero

intragenerational equity the concept that affluent groups in the community should be treated in the same manner as those less wealthy

investment analysis overall process for determining project viability prior to the decision to build and thereafter as an ongoing management input

investment return weighted mix of the expected interest, dividend or other rate of return on available equity per annum and the cost of borrowed funds per annum, for the purpose of assessing the investor's true time value of money

land cost purchase cost of the land

life-cost total cost of an asset measured over the period of financial interest of the owner (similar terms include *life cycle cost* and *cost-in-use*)

life-cost analysis that part of the cost control process dealing with the monitoring and recording of life-costs, the compilation of feedback and the management of future life-cost performance during construction and occupation

life-cost plan document that sets cost targets for the purpose of cost control by estimation of the total cost of a building project using an elemental approach (includes sketch design and tender document life-cost plans)

life-cost planning that part of the cost control process dealing with budgeting, optimisation and documentation of life-costs during design

life-cost studies the investigation of the cost of an asset using a total cost approach consisting of analysis, planning and evaluation activities

life cycle cost see *life-cost*

maintenance cost annual and intermittent costs associated with the repair of the building, including periodic replacement or planned renovation (may be conveniently divided into categories for repair and replacement)

management action plan initiatives for the performance improvement of buildings and components during construction and/or occupation

marginal utility the extra value obtained or obtainable from the acquisition of one more unit of consumption (or income)

measurement quantification of life-costs in real terms (see *present value*)

net present value the sum of the discounted present (or future) value of all expected cash inflows and outflows for a project over a selected time horizon

occupancy cost costs of staffing, manufacturing, management, supplies and the like that relate to the building's function, including denial-of-use costs

operating cost expenditure required to maintain the land and building and facilitate its function

opportunity ability of money to theoretically increase in value due to potential investment return on equity in real terms after tax

opportunity cost concept that money has investment value

ownership cost regular running costs such as cleaning, rates, electricity and gas charges, insurance, maintenance staffing, security and the like (may be conveniently divided into categories for cleaning, energy and other)

period of financial interest the period of time, expressed in years, during which the investor of the building is considered to have a 'financial' involvement

post-occupancy evaluation study of an occupied building so as to gain knowledge about its functional performance and to provide information for future design decisions

present value real value of costs and benefits to the present generation equal to the future cost expressed in constant dollars – used for measurement and control purposes and as the basis of the life-cost plan (not to be confused with the term *discounted present value*)

purchase cost acquisition costs not directly associated with the finished product, including items such as stamp duty, legal costs, building fees, professional fees, commissioning and the like

quality of life the level of attainment of general lifestyle objectives and prosperity (see *standard of living*)

real value representation of worth in constant dollars (see *present value*)

repair cost cost of maintaining a building component in a desired condition consistent with its age for the purpose of ensuring its proper performance (see *maintenance cost*)

replacement cost cost of demolition and reconstruction of building components that have reached a state of obsolescence (see *maintenance cost*)

residual life remaining component life at the end of the selected time horizon for the life-cost study

residual value theoretical worth of the residual life of a component, used in comparison of alternatives having different component lives

risk measure of the level of uncertainty of future events

risk analysis deterministic or probabilistic techniques for the identification and assessment of risk

selling cost expenses associated with ultimate sale, including real estate agent commissions, stamp duty, transfer fees and the like

sensitivity analysis a form of risk analysis comprising the testing of minor changes in key variables or assumptions to gain insight into the reliability of a decision

standard of living subset of the quality of life based on materialistic criteria measured at the level of the individual in society by expenditure on goods and services consumed over time in real terms per capita.

study period period of time covered by an economic appraisal or life-cost plan and recommended to be determined by the period of financial interest of the owner or investor

sub-element part of an element that is physically and dimensionally independent and separable in monetary terms

sustainability constraint theoretical cost or benefit of environmental damage or resource depletion expressed as an undiscounted annual equivalent and debited to investment projects over their study period

sustainable development the balance between economic progress and environmental conservation, comprising and acknowledging the concepts of environmental value, futurity and equity

target taxation rate control against which later performance can be compared rate at which income tax is paid on net earnings or profit per annum

time horizon see *study period*

time preference a philosophy of discounting variously based on pure impatience, risk and uncertainty and intergenerational equity

time value of money the concept that money changes over time, as a result of opportunity and time preference considerations

total cost approach the recognition that the real cost of an asset is the sum of all its initial and recurrent expenditures incurred over a specified time horizon

value worth of an object or service in terms of its function, desirability, cost and resale potential, encompassing both objective and subjective considerations and influenced by individual perception and expectations

value for money the effective balance between maximum function and minimum life-cost

value management structured and systematic technique for the identification of function and performance and its achievement at minimum life-cost (i.e. value for money)

value management study results of the value management technique applied to particular building components, elements or systems

RECOMMENDED READING

Ashworth, A (1994) *Cost Studies of Buildings* (Second Edition), Longman.

Boardman, AE (1996) *Cost-Benefit Analysis: Concepts and Practice*, Prentice Hall.

Boussabaine, A & Kirkman, R (2004) *Whole Life-Cycle Costing: Risk and Risk Responses*, Blackwell Publishing.

Bull, JW, (1992) *Life Cycle Costing for Construction*, Thomson Science & Professional.

Daly, HE (1997) *Beyond Growth: The Economics of Sustainable Development*, Beacon Printing.

Dell'Isola, AJ & Kirk, SJ (1995) *Life Cycle Cost Data* (2nd edition), McGraw-Hill.

Diesendorf, M & Hamilton, C (1997) *Human Ecology, Human Economy*, Allen & Unwin.

Fabrycky, WJ & Blanchard, BS (1991) *Life-Cycle Cost and Economic Analysis*, Prentice Hall.

Field, BC (1994) *Environmental Economics: An Introduction*, McGraw-Hill.

Flanagan, R & Norman, G (1983) *Life Cycle Costing for Construction*, Surveyors Publications.

Flanagan, R & Norman, G (1993) *Risk Management and Construction*, Blackwell Science.

Flanagan, R & Tate, B (1997) *Cost Management of Building Design*, Blackwell Science.

Flanagan, R, Norman, G, Meadows, J & Robinson, G (1989), *Life Cycle Costing: Theory and Practice*, BSP Professional Books.

Hanley, N, Shogren, JF & White, B (1996), *Environmental Economics in Theory and Practice*, Oxford University Press.

Goodstein, ES (1998), *Economics and the Environment* (Second Edition), Prentice Hall.

Kirk, SJ & Dell'Isola, AJ (1995), *Life Cycle Costing for Design Professionals* (Second Edition), McGraw-Hill.

Langston, C (1991) *The Measurement of Life-Costs*, NSW Department of Public Works.

Langston, C (1991) *Guidelines for Life-Cost Planning and Analysis of Buildings*, NSW Department of Public Works.

Langston, C (1994) 'The determination of equivalent value in life-cost studies: An intergenerational approach', PhD Dissertation, University of Technology, Sydney.

Langston, C (1996) 'Life-cost studies' in *Environment Design Guide*, Royal Australian Institute of Architects.

Langston, C & Ding, G (2001) *Sustainable Practices and the Built Environment*, Butterworth-Heinemann.

Pearce, DW, Markandya, A & Barbier, EB (1989) *Blueprint for a Green Economy*, Earthscan Publications.

Price, C (1993),*Time, Discounting and Value*, Blackwell Publishers.

Seeley, IH (1997) *Building Economics*, Macmillan Press.

Van Pelt, MJF (1993) *Ecological Sustainability and Project Appraisal*, Avebury.

INDEX